CURRICULUM and ASSESSMENT
for WORLD-CLASS SCHOOLS

HOW TO ORDER THIS BOOK

BY PHONE: 800-233-9936 or 717-291-5609, 8AM–5PM Eastern Time

BY FAX: 717-295-4538

BY MAIL: Order Department
Technomic Publishing Company, Inc.
851 New Holland Avenue, Box 3535
Lancaster, PA 17604, U.S.A.

BY CREDIT CARD: American Express, VISA, MasterCard

BY WWW SITE: http://www.techpub.com

PERMISSION TO PHOTOCOPY–POLICY STATEMENT

Authorization to photocopy items for internal or personal use, or the internal or personal use of specific clients, is granted by Technomic Publishing Co., Inc. provided that the base fee of US $3.00 per copy, plus US $ .25 per page is paid directly to Copyright Clearance Center, 222 Rosewood Drive, Danvers, MA 01923, USA. For those organizations that have been granted a photocopy license by CCC, a separate system of payment has been arranged. The fee code for users of the Transactional Reporting Service is 1-56676/97 $5.00 + $ .25.

THE SCHOOL LEADER'S LIBRARY: LEADING FOR LEARNING

CURRICULUM and ASSESSMENT for WORLD-CLASS SCHOOLS

BETTY E. STEFFY, Ed.D.
Dean of the School of Education
Indiana University/Purdue University at Fort Wayne

FENWICK W. ENGLISH, Ph.D.
Vice Chancellor of Academic Affairs
Indiana University/Purdue University at Fort Wayne

SERIES EDITOR: PAULA M. SHORT

TECHNOMIC
PUBLISHING CO., INC
LANCASTER · BASEL

Curriculum and Assessment for World-Class Schools
a TECHNOMIC publication

Published in the Western Hemisphere by
Technomic Publishing Company, Inc.
851 New Holland Avenue, Box 3535
Lancaster, Pennsylvania 17604 U.S.A.

Distributed in the Rest of the World by
Technomic Publishing AG
Missionsstrasse 44
CH-4055 Basel, Switzerland

Copyright ©1997 by Technomic Publishing Company, Inc.
All rights reserved

No part of this publication may be reproduced, stored in a
retrieval system, or transmitted, in any form or by any means,
electronic, mechanical, photocopying, recording, or otherwise,
without the prior written permission of the publisher.

Printed in the United States of America
10   9   8   7   6   5   4   3   2   1

Main entry under title:
   The School Leader's Library: Leading for Learning
     Curriculum and Assessment for World-Class Schools

A Technomic Publishing Company book
Bibliography: p.
Includes index p. 143

Library of Congress Catalog Card No. 96-61339
ISBN No. 1-56676-438-6

TABLE OF CONTENTS

Preface to the Series   ix

Preface   xi

Chapter 1. Fundamental Curriculum Concepts 1
    Curriculum Quality Control   4
    The Plurality and Layering of Curriculum   4
    The Socioeconomic Determinism of
        Unaligned Testing   5
    Curriculum Design and Delivery   7
    The Difference between Achievement
        and Learning   8
    The Knotty Issue of Teaching to the Test   9
    Two Kinds of Alignment   11
    Getting a Handle on the Taught Curriculum   12
    Curricular Metaphors   14
    The National Debates over a Curriculum Core   16
    References   18

Chapter 2. Curriculum Planning 21
    The Matter of Morals   23
    Where to Start the Planning Process   24
    Curriculum as the Means for Nation Building   31
    References   33

Chapter 3. World-Class Curriculum 35
    Role of National Standards   36

The Fight for Academic Instructional Time   44
Standard Setting at the State Level   44
What Other Countries Do   48
Opportunity to Learn   51
What the Graduating Senior Is Expected to Know and Be Able to Do   53
A Paradigm Shift   59
New Standards Project: Linking Standards to Assessment   62
References   64

Chapter 4. Toward Balanced Assessment 67

Rethinking Assessment   67
State Assessment Practices   68
Achievement as Measured Learning   70
Example: Portfolio Assessment   71
Example: Open-Ended Response Item   76
Example: Performance Event   80
Authentic Assessment as a Catalyst for School Reform   84
International High School   85
The Bronx New School   92
Using Assessment to Transform Schools   95
References   96

Chapter 5. Program Evaluation 97

Program Evaluation Standards   98
Guidelines and Common Errors for Selected Standards   99
Limitations of Program Evaluation   104
Types of Program Evaluation Designs   105
Sources of Resistance to Program Evaluation   108
Developing Program Evaluation Designs   111
Appendix   118
References   121

Chapter 6. Toward Continuous Curricular Improvement 123

National Priorities   123

State Initiatives   125
Foundation Funding   126
Professional Based Initiatives   126
Local Board Initiatives   126
Curriculum Development as
   Political Consensus   127
International Standards and Tests   129
National Failures to Improve Test Scores   130
Battles over Privatization and Vouchers   130
Creating a Systemic Response   131
References   135

Chapter 7. A Keyhole Peek into the 21st Century **137**
The Curriculum Challenges Ahead   138
Closing Thought   140
References   141

Index   143

PREFACE TO THE SERIES

SCHOOL leaders must constantly search for opportunities to improve the learning opportunities for students. Managing the complex issues that characterize the work of school leaders requires that they deal with legal questions, curriculum change, student assessment, staff development, and other issues. School leaders also are faced with a rapidly changing environment characterized by new technology, diversity of communities, a focus on global understanding, and multiple school reform initiatives. Each compete for the attention and expertise of the school leader. New instructional strategies and new roles for school personnel demand that the school leader have ready access to information that informs the decision-making process.

The book series, *The School Leader's Library: Leading for Learning,* provides the busy school leader with critical information to assist in understanding the complex issues impacting schools. Each book in the series addresses a critical concern of school leaders in a format that presents current, useful information. The books also are useful to students in leader preparation programs and professors who teach in these programs. While practical in approach, each book contains a thorough review of relevant literature that informs the key issues. The authors are noted experts in the particular topics presented in the books in the series. These books belong in the school leader's library as well as university classrooms.

PAULA M. SHORT
Series Editor

PREFACE

IF one ever believes the idea that a consideration of curriculum is dull and boring, initiate a conversation with any American over what children should be learning in schools. The chances are quite good that you will receive a rather definitive list of topics or skills. Unlike some aspects of education that have become the domain of specialists such as testing and assessment, curriculum as an idea of content to be taught remains in the concrete and lived experiences of most Americans who have been to school.

This small book about curriculum in American schools is centered on answering the questions regarding placing it within the context of schools and school systems (organizational relationships) and presenting ideas regarding world-class standards and continuous improvement. It is not the result of armchair speculation, but the product of interaction and discussion with practitioners at all levels of public education in the United States, largely in the presentation of curriculum alignment and curriculum auditing training. Above all else, a discussion about curriculum in schools assumes emotional overtones largely connected to high-stakes assessment and the necessity to show that all children are learning.

We have not neglected theoretical matters however. Perhaps the most important of those is that curriculum cannot be allowed to become simply the means to continue to separate the haves from the have-nots based on socioeconomic status (SES) alone. Both coauthors recoil at the practices of some of the states to simply engage in standardized testing of pupils that separates them into socioeconomic strata with alarming reliability. When SES is attached to matters of race and gender, we have witnessed pseudo-scientific racism at work. When state officials insist that these measures are stable indicators of intelligence rather than wealth derivative, we counter with the idea in accountability that one has to be in control of one's fate to be accountable. If test performance is based largely on unchanging and heritable IQ, to what extent can school administrators and teachers be accountable?

Likewise, if achievement is predicted by SES, how accountable can one be for continued low pupil performance? We reject both genetic and environmental determinism. We consider both to be interactive and alterable. The issue is alignment—teaching what one tests.

To this end, we eschew tests that are secretive, unaligned, and dubious measures of pupil learning when they are not related to what students have been taught in schools. There can be no accountability without alignment. We also call into question the seductive idea that reform is really about nothing more than "freeing the schools" to go their own way. Particularly secondary schools become big losers in any accountability plan that is backed with high-stakes assessment. School learning is complex and continuous. It is linked by the presence of a curriculum. It is only with a linked vertical curriculum that a test is a fair measure of what students have learned in secondary schools. The need for focused continuous learning requires a curriculum, and it requires control and connection of the elementary and middle schools to the high school. Otherwise, in the same school district, we have witnessed successful elementary schools and unsuccessful secondary schools. What we are witnessing is suboptimization, the situation where one part of the system is successful at the expense of the remainder of the system.

World-class schools require comparison to world standards. The intense localism of educators and citizens regarding a discussion of national educational goals dooms American schools to a second-class status. It means that American students will continue to be tested on that which they have not been taught, placing them in a most unfortunate position compared to their global student counterparts with national consensus regarding their educational systems complete with a national curriculum.

We doubt that Americans can compete well in the 21st century with a curriculum defined by peculiarly narrow boundaries and interests. If we want American schools to be "world-class," we must be willing to accept some common national and even international standards for our schools. Otherwise, we have no rational means to make any claims about how good our schools are. The world does not run on Yankee time.

We want to thank our colleagues in the trenches all over the world who work every day to make a difference. That they succeed more often than they fail is in spite of rather than because of, the curriculum. We hope that this book will make their work more rewarding, as well as more effective for their students.

CHAPTER 1

Fundamental Curriculum Concepts

CURRICULUM is the work plan designed for teachers to reference or follow as they define their teaching. Teaching can exist in the absence of any work plan. It can occur without curriculum guides, textbooks, or any supportive materials. In the end, teachers can teach what they know.

When it becomes critical for teachers to engage students in learning that has been predefined as important or crucial, and that learning is complex requiring a multi-year classroom focus in which more than one teacher is involved, a work plan becomes the means to focus and connect all of the teachers toward priority ends. It is the necessity for focusing and connecting teacher work that curriculum as a work plan comes into being. The function of curriculum in schools is to establish the framework for purposive teaching, hereinafter referred to as instruction. Curriculum is the intervening variable between teaching as a random act and instruction, which is a preplanned, anticipated act. Instruction is focused and connected teaching. Instruction is the major internality and reason for curriculum in schools.

The development of curriculum can occur without any reference to externalities such as textbooks, tests, national goals or standards, or the condition of the world. However, if curriculum is to be used to improve desired learning by directing teaching, then it must be aligned to that externality. Curriculum can be developed without alignment to anything external; that is, it can create greater focus and connectivity and not be aligned. In this case curriculum can never be the vehicle to improve a test score, attain mastery of a book, or meet world-class standards unless these are embedded in the curriculum as it was created.

Figure 1.1 shows these relationships in schematic form. We know of one large urban school system that spent over a million dollars drafting a curriculum that was not referenced to any externality such as the state basic skills test or world-class standards. In this case if the curriculum were followed, it would create focus and connectivity, but since it was not aligned or matched to any externality, it could not be used to improve performance as measured on that which is related to a test or standards. Working harder in this situation is futile as far as improving test/assessment performance is concerned.

Two aspects of focus and connectivity are important benchmarks for instruction. One is the vertical manifestation called articulation. The second is referred to as coordination. Curriculum coordination is created when teachers at some similar place in the school or system focus on similar things or processes to teach, for example, if eighth grade math teachers agree on common concepts and skills to teach in the year.

Articulation is concerned with focus and connectivity from grade-to-grade, level-to-level, or school-to-school. It is possible to create curriculum coordination without dealing with articulation. On the other hand, articulation assumes coordination is present.

The purpose of curriculum as a work plan is to influence the

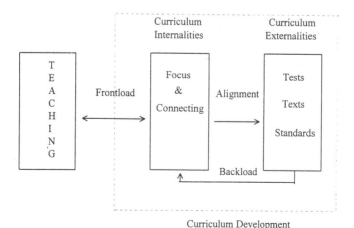

Figure 1.1. *Using curriculum to improve teacher effectiveness.*

work teachers are expected to perform. Curriculum must exist within a contrived human organization known as a school. A school is a special kind of work place. It is an architectural form built around four walled spaces connected to long passageways. The spaces are most often uniform. Unless a class is designed as some special part of the curriculum, such as a shop or home economics lab, it is impossible to tell when examining the empty space to know if the cubicle contained a teacher of history, English, algebra, Spanish, or journalism.

Schools contain a good deal of uniformity inside. Curriculum is supposed to "fit" into this structure rather than the structure designed to match the curriculum. Curriculum is a common fluid filling water glasses. If the water is clear and the glasses standardized, it is impossible to differentiate among them by sight.

Curriculum may differ by content area and by sophistication such as Advanced Placement, middle level English, Honors Algebra, or dumbbell composition. In most other ways curriculum does not differentiate or impact the structure of the school. It is designed not to interfere with that structure, that is, time/space manifestations. In this case one can speak of curricula which differentiates between response to teaching (fast, slow) or to content depth (advanced English) or to teaching method (inductive, deductive), but not to issues of radical unevenness in those same manifestations. The type of school schedule is perhaps the most important decision an administrator can make about curriculum where the curriculum follows the school structure rather than leading it.

Current school structure is designed to promote orderly groups of students ready for didactic teaching. This reality pervades one of the tacit goals of curriculum construction, sometimes called the "hidden curriculum" (English and Hill, 1994, p. 16). The "hidden curriculum" is also called the "lived curriculum" or the implicit socio-political relations which make up the "structured silences" of schools (see Aronowitz and Giroux, 1985, p. 75; English, 1988, pp. 89–105; English, 1991, pp. 84–104).

Figure 1.1 also references two processes of creating alignment: frontloading, that is, developing curriculum first and designing/adapting/purchasing tests second, or backloading, beginning with the test first and using it as the basis for creating a curriculum (see English, 1992, pp. 83–86). In the former case, alignment is more problematic than the second. However, in the second case (backloading) issues of teaching to the test are raised which are tendentious and not well understood by the general public. They are often confused by educators themselves in practice.

CURRICULUM QUALITY CONTROL

Quality control refers to the internality of the capability of an organization to connect its operations together and to be responsive to external cues (English, 1987b, p. 101). There are three generic curricular contents that provide directions to teachers inside of schools and school systems. They are the written curriculum, the taught curriculum, and the tested curriculum (English, 1987b). Successful curriculum management is considered bringing these three types of contents into congruence or focus and connecting them. The result that emerges is that more of the curriculum that is tested is taught. When this occurs, student test performance is improved.

One of the chief deterrents to this convergence is that school systems only appear to be internally integrated. In reality they are layered and only superficially connected. As different layers they communicate laterally but not vertically. Thus, the board of education may adopt goals or rule on other curricular matters. Teachers often miss such directives (or ignore them) altogether. The presence of board-adopted goals at one layer does not mean they influence any other layer, except that the layers touch one another like a cake. The phenomenon of the "layer cake" has been called by one researcher "loose coupling" (Weick, 1978). This is also why educational reforms of school governance rarely change teacher behavior in classrooms (see Quinn, Stewart, and Nowakowski, 1993).

THE PLURALITY AND LAYERING OF CURRICULUM

There is another reason for the difficulty of managing curriculum in schools beyond the layer cake phenomenon. This reason has to do with the fact that curriculum is more than curriculum guides. Curriculum is any document a teacher may reference in the act of deciding what to teach, when to teach it, how much of it to teach, and perhaps the method of teaching itself.

Historically, there is a plethora of paper locked into the classroom teacher's cabinet. These may consist of textbooks, teacher guides to textbooks, accreditation guidelines, scope and sequence charts, curriculum guides, state department guidelines, board policies, the principal's directives, the supervisor's recommendations and the like. Curriculum auditing experience shows that rarely are all these pieces of paper internally aligned with one another. They are often contradictory, sometimes providing the teacher with conflicting recommendations (English, 1988, p. 53).

The reality of classroom life is that the teacher is most often alone with students. When the door is shut the teacher determines what piece of paper (if any) he/she will follow to shape his or her work. The presence of a plethora of paper in schools also creates a kind of insouciant cynicism among teachers about their value. Too often, locally developed curriculum is connected to dominant administrative or board of education personalities and disappears when they exit the school system in the electoral process. For this reason the "real" taught curriculum may float somewhere between the officially adopted written curriculum and the imposing reality of the tested curriculum.

THE SOCIOECONOMIC DETERMINISM OF UNALIGNED TESTING

The tested curriculum refers to the actual content of the test, that is, the "stuff" around which the test questions revolve. It could be facts, processes, applications, problem-solving capabilities and be scored by machine or by humans using a scoring rubric. Without alignment, tests become measures of student socioeconomic status or SES. For example, scores on the NAEP (National Assessment of Educational Progress) were found to be SES related. Eighty-nine percent of the variance of the scores was explained by four variables: the number of parents living at home, the parents' education, community type, and the state poverty rate (see Robinson and Brandon, 1994). In most states, the highest test scores emanate from the richest school districts and the lowest from the poorest. These are often associated with race because of the historic discrimination directed against minorities and their continuing inability to advance economically within the larger society. The extremes between rich and poor are the greatest in the United States of any nation in the world. Whereas in Britain 1 percent of the population owns about 18 percent of the nation's wealth, in the U.S., 1 percent owns 40 percent of the country's wealth (Bradsher, 1995, p. C4) and the gap between the haves and the have-nots is growing. Tests reflect wealth disparity, particularly so if they assess the "cultural capital" that money buys, as opposed to the actual taught school curriculum. On unaligned tests, no school-related variable predicts statistically significant scores. The major predictors are socioeconomic, a fact first reported in the 1960s by the late James Coleman (1966).

Those persons in the bottom income ranks of U.S. society have seen their real annual earnings drop by 24 percent since 1973, while the top 20 percent increased by 10 percent in the same time period. Between 1973 and 1993, males with only a high school education lost 30 percent of their real income (Rattner, 1995, p. A19). SES also is interactive with teacher expectations and becomes converted to academic achievement. Robinson (1994) reported in a study of a South Korean elementary school that teachers expected more of higher SES students than lower ones. Furthermore, teachers exerted tighter control over lower SES students and called upon them less frequently than higher status students. School did not act as an ameliorative agency in enabling poor students to learn more. School "reproduced" the social class system instead of changing it. The concept of SES reproduction in the schools has been called "correspondence theory" by Bowles and Gintis (1976); that is, school "corresponds" to the society and reproduces it largely intact. Unaligned tests simply cement the correspondence, especially for low SES students.

CURRICULUM DESIGN AND DELIVERY

Two facets of curriculum generally preoccupy school practitioners. The first revolves around issues pertaining to the physical creation of a curriculum, that is, a work plan to guide teaching. These focus on curriculum design. The second is concerned with implementation of the plan once developed, that is, curriculum delivery (English and Steffy, 1983).

Unraveling curriculum problems rests on understanding whether it is a design or a delivery problem and how one influences the other in school settings. For example, in one southern state in the mid-1980s, the state department converted an eighth grade math test to an eleventh grade exit test of math competencies. When droves of secondary school students failed the test, there was a mad scramble to understand what happened.

If one believes a low test score is a delivery problem, then one may ask, "What's wrong with teachers? Why aren't they teaching students properly?" That question ignores the fact that there may be something wrong with the curriculum, which would be a design problem. "Fixing" a problem means understanding its source. A design problem is fixed by correcting its alignment. A delivery problem usually involves more effective supervision and/or staff development.

In the case of the southern state, it was determined that teachers were using the adopted state text as they were required to do by law. The majority were doing so successfully. An examination was then conducted of the extent to which the skills of the math test were actually included in the adopted state text. When it was determined that the alignment was low (less than 50 percent), the problem would have to be solved by increasing the match between the adopted state text and the state's test. This is a design problem. It is "fixed" by improving the match between the written and tested curricula.

As it turned out, a large majority of the principals in the same state incorrectly read the low scores as a delivery problem. They apparently reasoned that the test and text were adopted by the state agency, which meant they had to be aligned. As teachers were using more than the single state text (multi-text), they had slightly improved the low alignment. However, principals saw the multi-text approach as "detracting" from the alignment. Many ordered their teachers to drop the other books and concentrate on the state's text. When teachers did so they lost the alignment and the test scores dropped again.

Many schools and school systems have design and delivery problems simultaneously. In such cases, they should correct their design (curriculum/test match) prior to concentrating on delivery issues. If the curriculum is not aligned with the test(s), improving curriculum delivery will not occur until the test/curriculum content is mastered and more of it is included and taught.

THE DIFFERENCE BETWEEN ACHIEVEMENT AND LEARNING

Improving pupil achievement means knowing what to concentrate on in classrooms to maximize alignment. That means knowing the difference between achievement and learning. Put succinctly, achievement is only the learning that is tested. Not all learning is tested. In this case, one can learn more but look bad on a test if the test doesn't match what one has learned.

Many administrators and classroom teachers have great difficulty understanding that an improvement in test scores occurs when one concentrates on achievement. That means knowing the difference between the tested curriculum and the curriculum that isn't tested. To improve test scores one doesn't simply teach more, one teaches more of what is tested. Grasping this differentiation means that educators would examine test specifications carefully, search for sample test questions to understand format variations, and attempt to understand the frequency of content, skills, or processes tested. Are there fourteen items about percentage computation and only two about decimal fractions on the test?

Some state department administrators in one state were fond of saying that there should be a "seamless web" between the written, taught, and tested curricula. We do not support this concept because it would mean that both the written and

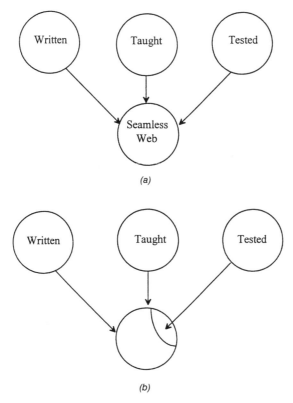

Figure 1.2. (a) The congruence of "the seamless web"; (b) making sure achievement is not left to chance.

the taught curricula would be limited to only that which was tested. Figure 1.2 illustrates the difference. In the first drawing the written, taught, and tested curricula are shown coming into total congruence. This means that there would be no difference between the three. In the second instance, the test is included in the taught curriculum, but there is more taught than tested.

THE KNOTTY ISSUE OF TEACHING TO THE TEST

There are few more troubling areas concerning alignment than the one of "teaching to the test." Adopting the general rule that one doesn't teach to the test is as rigid and incorrect as assuming that one always teaches to the test.

The educator must take into account the kind and purpose of testing as well as the consequences under an accountability scheme. These issues are, unfortunately, mixed within legislatures and state departments of education. There is great variation among state agencies on this question as well as accompanying rules and practices.

Our rules are straightforward and simple. If the purpose of the test is diagnostic and is not to be used as a tool to distribute rewards and punishments under some accountability scheme, one would not teach to that test. Diagnostic tests may or may not be aligned to a local curricula. If, on the other hand, the test is to determine the extent of pupil learning for accountability purposes, it should be matched to the curriculum or accountability is meaningless. Without alignment, most tests simply mirror the SES levels of the largest number of children from the dominant level served by a school or school district. High SES schools score well, low SES schools do poorer. And no amount of punishment will change a thing until the curriculum and test are aligned.

That this is correct underscores the reason why when tests are kept a secret to insure a so-called normal curve response, school administrators are urged to compare their district's scores to systems "like" theirs. This translates to similar SES compacted variables. A suburban Houston school district would not compare its scores to inner-city Detroit, but to a system "like" itself as far as pupil population and SES were concerned. The reason is that these are the only variables that predict the results obtained.

A state agency that insists on using a secret, diagnostic, unaligned test as an accountability tool to reward and punish administrators and teachers is simply punishing low SES students and the professionals who serve them. In such circumstances, there are no "poor schools." There are only "schools who serve the poor."

When parents want standardized norm-referenced tests administered to their children to see "how our kids are doing compared to others nationally," all they are comparing is one set of students with an unaligned (random) curriculum to other similar SES kids someplace else with their own unaligned curriculum. In neither case was the test a measure of a specific local curriculum, and therefore it can't be a measure to determine "quality education." It is a measure of a random variable, particularly if the normal curve is applied to results. Tests that are kept a secret cannot be tools to determine quality education or accountability. All accountability tools assume that those who are to be accountable control the variables that determine their success or failure. This is not the case with secrecy or the application of artificial measures of scarcity (the bell-shaped curve). Randomness and accountability are oil and water. They don't mix.

One does not teach to a test if it is a diagnostic instrument chiefly designed to determine capability, readiness, or responses that have been designed to relate to measures of randomness. The bell-shaped curve is not the frequency distribution to judge school learning. That is the function of a "J" curve (English, 1992, p. 74).

State agencies are notorious for mixing purpose and tests and for leaving local educators with the Hobson's choice to be ranked low because of the presence of large numbers of poor SES students in their schools or to be castigated and punished for "teaching to the test," that is, cheating. Without alignment for school systems serving large numbers of low SES children, there is no way off the bottom of the imposed bell curve. In such situations accountability is a cruel joke.

TWO KINDS OF ALIGNMENT

Broadly speaking, there are two kinds of alignment shown in Figure 1.3. Content alignment refers to the match between the body of knowledge, skill, process in the written curriculum and that within the test. It is a design issue. Context alignment refers to the match between the format of the curriculum and its match to test format. This is also a design issue.

Delivery issues are concerned with the match between what the teacher taught that was included in the curriculum and

	Content Alignment	Content Alignment
Design Alignment	Test/Curriculum Content Match	Test Format Matches Curriculum Format (Without Item Specifics)
Delivery Alignment	Content of Curriculum is Taught as Designed	Teacher Includes Format Response in Learning Environment

Figure 1.3. Two kinds of alignment in two different scenarios.

whether or not the teacher placed the format of the test items into the learning environment in order to practice what the context required. Item specifics would be excluded. This means that in a particular type of word problem, the actual numbers or specifications would not be the same from the practiced problem to the tested problem.

Alignment is a form of transfer theory which was formulated by E. L. Thorndike at the turn of the century (English, 1987b, pp. 158–161). Essentially, transfer theory posits that when one wants to improve the probability that something learned in one situation will be transferred to another, one makes the two situations as identical as possible. This principle is nearly universally applied in special education, vocational education, and athletics in school curricula. The largest scale use of the principle occurs in U.S. military training.

GETTING A HANDLE ON THE TAUGHT CURRICULUM

Part of managing curriculum successfully is developing an approximation of the content really being taught by teachers in classrooms. The practice of asking teachers what they actually taught is curriculum mapping (English and Steffy, 1983b).

Curriculum maps can be three dimensions. A one-dimensional map records curriculum content only. A two-dimensional map records content and time spent. A three-dimensional map is a record of content, time and sequence. A two-dimensional map is shown in Table 1.1 (English and Steffy, 1983b, p. 20). Fluctuations occur in time spent because of pupil mastery requirements, different textbook emphasis if one textbook is not in use, or teacher interest that can be independent of all other variables. Some teachers "don't like" to teach some things.

Curriculum maps simply indicate at the outset that wide variations may exist for a variety of reasons in the same school system. For example, in Table 1.1 Junior High School D spent 9.44 class periods with the "research paper." By contrast, Junior High School E spent less than a full period on the same area. Unless Junior High School E is dealing with accelerated pupils or ones who have nearly mastered "the research paper" (or it is taught somewhere else in the curriculum), it is doubtful that the variation is due to pupil needs and more toward teacher interest. The same most likely would be the case with the objective pertaining to the "essay." Junior High School C spent no time on this topic. This "curriculum hole" is very likely a demonstration of teacher inattention. Large time differences on maps such as these are usually indicative of lax supervision and teachers straying from adopted curricular guidelines

Table 1.1. Two-dimensional curriculum map showing average time spent by class periods on strands/objectives for English 9 among six junior high schools.

Objectives	District Average	JHS A	JHS B	JHS C	JHS D	JHS E	JHS F
1. Five Paragraph Essay	6.15	10.72	7.67	1.65	3.28	8.11	8.04
2. Research Paper	5.64	6.69	6.92	1.30	9.44	0.07	11.00
3. Phrases-clauses	2.96	4.00	7.17	0.45	3.11	2.04	2.42
4. Main Idea	2.45	2.13	4.08	0.75	2.47	5.75	0.17
5. Outline	2.42	6.16	2.75	0.20	3.00	1.07	1.50
6. Sentence Types	2.25	3.06	4.79	0.80	1.11	0.25	5.08
7. Essay	2.24	3.22	3.71	0.00	1.33	2.36	4.46
8. Implied Meaning	2.18	1.03	4.17	0.35	2.86	4.39	1.21
9. Reading Flexibility	2.12	5.34	0.33	1.05	3.94	0.96	0.00
10. Types of Literature	1.66	1.84	5.67	0.05	1.50	1.25	0.83
11. Cause/Effect	1.63	2.53	1.04	0.55	1.53	4.14	0.00
12. Meaning/Context	1.61	1.13	4.42	0.43	1.00	2.79	0.96
13. Speech	1.13	0.00	0.00	2.55	0.44	2.57	0.75
14. Take Notes	0.68	0.66	0.58	1.30	0.39	0.75	0.13

based on their own interests rather than being a reflection of differences in learner style or ability. Variations in time are expected in curriculum maps. If there are no variations, it is impossible to claim the curriculum is being tailored to pupil need, interest or ability. Uniformity would indicate that the curriculum was being delivered (taught) rigidly, without any input from students.

Curriculum maps are valuable sources for revealing what is actually going on in classrooms as opposed to trying to discern the data by observation, which is too time consuming, or reviewing teacher plan books. Maps can be simply constructed and even anonymously completed. Both uniformity and variation are important trends in mapping the taught curriculum.

CURRICULAR METAPHORS

A metaphor is a kind of mental image carried in the brain about the meaning of a concept. When discussing curriculum one thinks of either a staircase as shown in Figure 1.4 or a spider web as illustrated in Figure 1.5. These two metaphors sym-

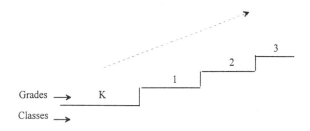

Major Characteristics

- Linear and segmented
- Fragmented
- Logically arranged by perceived level of difficulty (scope and sequence)
- Easily scheduled and evaluated (testing)
- "Balance" is determined prior to implementation, prior to learning occurring or even teaching; subject matter is content-dominated; whatever "adjusting" occurs is a matter of proper motivation and pacing *within and not outside the model*
- Event and sequence accountability are established and maintained

Figure 1.4. *The traditional metaphor of curriculum—the staircase.*

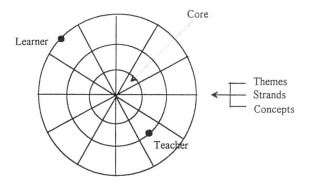

Major Characteristics

- Nonlinear
- Cohesive, holistic
- Difficult to standardize and evaluate
- Difficult to schedule
- More psychological than logical
- Balance determined by simple *inclusion* decisions rather than by scope and sequence decisions
- Learner very active in making "sense" of the curriculum
- Teacher more of a partner than controlling agent of the curriculum
- Nature of "accountability" changes; it doesn't matter *when something* is learned, and specific facts don't count

Figure 1.5. The spider web metaphor of curriculum.

bolize two different ideas of the nature of curriculum. The first is the traditional. It is anchored in the idea of proceeding from simple to complex, small to large. One hears the idea that a child crawls before he/she walks. So they must know their "facts" before they can engage in more sophisticated endeavors. This idea is rooted in the psychology of B. F. Skinner and behaviorism. From this ideology, highly rational models of curriculum use behavioral objectives, educational outcomes, and logical ways to arrange and sequence a curriculum. Curriculum designers try to pre-specify nearly everything that a student should be able to do or know when encountering a curriculum. Measurement is impacted as well. Tests reflect the dominant psychology and curricular ideology in use. This model of curriculum is neat. It "fits" into the structure of education as it exists (see Apple, 1979).

The alternative metaphor is the spider web. It posits a very different idea of learning. Rooted in the psychology of Piaget (1950) and the linguistic research of Noam Chomsky, it believes in submersing the student in a complex process without fear. In the spider web everything cannot be pre-specified exactly because space must be provided for the learner to engage the curriculum and find personal meaning in his/her experience.

The approach has been called "constructivism" (Glatthorn, 1994). Constructivism is focused on broad-based, thematic approaches that are amenable to authentic assessment forms toward measuring learning. The constructivist focuses on individual differences between students as important and not a barrier toward schooling. The constructivist is not so concerned with identifying all of the possible results of an educational experience, but may use Eisner's (1969) "expressive objectives" instead of behavioral objectives. The former is concerned about student encounters as opposed to student outcomes.

Perhaps the most obvious example of a constructivist perspective in curriculum would be the "whole-language" approach to teaching reading. In this "spider web" model, language skills are integrated around common experiences and not taught in isolation one from the other. The linkage between reading and writing is stressed with the interconnectivity by which children have learned the spoken language of their culture.

THE NATIONAL DEBATES OVER A CURRICULUM CORE

The idea that there is or should be a "common core" of curriculum knowledge central to the education of all Americans is a point advanced by both educationists from the political right and left. Both assume that for a "core" to exist at all, learners have something important in common in order for a "core" to exist. If learners are so different that any commonality among them is non-existent or trivial, a "core" is absurd. It would also be illogical to talk about schools and classes as we currently understand them. A curriculum can only exist if it is assumed that learners have important commonalities that enable schools to group students for teachers to engage them in a core.

Not all psychologists concede this point. For example, Carl Rogers (1961) once expressed the view that, "It seems to me that anything that can be taught to another is relatively inconsequential, and has little or no significant influence on behavior" (p. 276). Rogers underscored the significance that only self-appropriated learning changes behavior and that such learning "cannot be directly communicated to another" (p. 276). The logical extrapolation of this view is that, "I have come to feel that the outcomes of teaching are either unimportant or hurtful" (p. 276).

If Rogers is correct nearly all discussions about curriculum, current notions of teaching and schooling, are "unimportant or hurtful." There is certainly evidence that for some children Rogers' perspective is accurate (see "America's Blacks," 1991, pp. 17–21). A call for vouchers and schools of choice will not correct these deficiencies.

Another view is that children are both similar and different. They are similar in that they are expected to take part in a common culture and to be supportive of the democratic processes of an elected form of government. The requirement for citizenship establishes the basis of a social contract for curricular commonality. Thus, the similarity is created by political necessity rather than psychology. This is the view of one of the most famous "core" advocates named Mortimer Adler (1982).

Adler's political rationale is borrowed by other writers and groups. For example, when defining what all Americans should know about science in Project 2061, The American Association for the Advancement of Science (1989) supported a common curriculum because, "the most serious problems that humans now face are global: unchecked population growth ... acid rain, the shrinking of tropical rain forests ... disease, social strife, the extreme inequities in the distribution of the earth's wealth ..." (p. 12). Since all nations of the world are involved in these global problems, a global curriculum encompassing them is viewed as the solution in the form of a common curriculum.

Most educators concede that, while children are indeed different, they do have common needs, exhibit common growth patterns, and can be grouped successfully for teaching to be accomplished. The extent of individualization involved depends upon the nature of the psychological and cultural differences that may exist among them at any given moment.

The most nettlesome dilemma with a curriculum core is what to include and exclude. Charges of bias have been raised in the suggestions of E. L. Hirsch's (1988) notion of "cultural literacy," that there is a set of facts that produce a culturally literate person. "Because culture is not a kind of capital that can be stored in a bank, one cannot 'add up the attributes of a culture and produce an individual'" (Pitman, Eiskovits, and Dobbert, 1989, p. 7).

Most Americans and most educators assume that some form of common experience in schools is essential to not only nation building, but to social learning that forms the cohesion among a nation's peoples. If this were not so, the federal courts would not have seen segregation as harmful not only to African Americans (those excluded) but to the whole fabric of the United States (those included). Segregation works both ways to be deleterious to all parties.

The discussion about a curriculum core is no more apparent than determining what a "world-class school" is. For any dialogue at all begins with determining what criteria may be in common by which "world-class" can be applied (see English, 1994, pp. 1–6).

As standards are discussed in ensuing chapters in this book, it should be remembered that the only way standards are relevant to guide school curriculum development is to concede that learners have something important in common. We believe that what is in common is a future in which all peoples globally can live productively, peacefully, and environmentally harmoniously.

REFERENCES

Adler, M. (1982). *The paideia proposal.* New York: Macmillan Publishing Company, Inc.

American Association for the Advancement of Science. (1989). *Project 2061.* Washington, D.C. AAAS.

References

America's Blacks. (1991). *The Economist*, 318(7700):17–21.

Apple, M. (1979). *Ideology and curriculum.* Boston: Routledge and Kegan Paul.

Aronowitz, S. and Giroux, H. A. (1985). *Education under siege.* South Hadley, MA: Bergin and Garvey.

Bowles, S. and Gintis, H. (1976). *Schooling in capitalist America.* New York: Basic Books.

Bradsher, K. (1995, April 17). Gap in wealth in U.S. called widest in west. *New York Times,* pp. A1–C-4.

Coleman, J. S. (1966). *Equality of educational opportunity.* Washington, D.C. Office of Education.

Eisner, E. (1969). Instructional and expressive objectives: Their formulation and use in curriculum. In R. E. Stake (Ed.) *Instructional objectives* (pp. 1–31) Chicago: Rand McNally.

English, F. W. (1987a). The principal as master architect of curricular unity. *The Bulletin,* 71(498):35–42.

English, F. W. (1987b). *Curriculum management for schools, colleges, business.* Springfield, Illinois: Charles C Thomas Publishers.

English, F. W. (1988). *Curriculum auditing.* Lancaster, PA: Technomic Publishing Co., Inc.

English, F. W. (1991). Visual traces in schools and the reproduction of social inequities. In Borman, K. M., Swami, P. and Wagstaff, L. P. (Eds.) *Contemporary issues in U.S. education.* Norwood, New Jersey: Ablex Publishing Corporation, 84-104.

English, F. W. (1992). *Deciding what to teach and test.* Newbury Park, CA: Corwin Press.

English, F. W. (1994). Foreword. In Jenkins, J. M., Louis, K. S., Walberg, H. J. and Keefe, J. W. *World class schools: An evolving concept.* Reston, VA: NASSP.

English, F. W. and Hill, J. C. (1994). *Total quality education: Transforming schools into learning places.* Thousand Oaks, CA: Corwin Press.

English, F. W. and Steffy, B. E. (1983a, February). Differentiating between design and delivery problems in achieving quality control in school curriculum management. *Educational Technology,* 13:29–32.

English, F. W. and Steffy, B. E. (1983b, Fall) Curriculum mapping: An aid to school curriculum management. *ERS Spectrum,* 1(3):17–26.

Glatthorn, A. (1994, October). Constructivism: Implications for curriculum. *International Journal of Educational Reform,* 3(4):449–455.

Hirsch, E. L.(1988). *Cultural literacy.* New York: Basic Books.

Piaget, J. (1950). *The psychology of intelligence.* New York: Harcourt Brace Jovanovich.

Pitman, M. A., Eiskovits, R. A. and Dobbert, M. L. (1989). *Culture acquisition.* New York: Praeger.

Quinn, D. W., Stewart, M. and Nowakowski, J. (1993, January). An external evaluation of systemwide school reform in Chicago. *International Journal of Educational Reform.* 2(1):2–11.

Rattner, S. (1995, May 23). *Wall Street Journal,* p. A19.

Robinson, G. E. and Brandon, D. P. (1994). NAEP test scores: Should they be used to compare and rank states? *Educational Quality.* Arlington, VA: Educational Research Service.

Robinson, J. (1994, November). Social status and academic success in South Korea. *Comparative Education Review,* 38(4):506–530.

Rogers, C. (1961). *On becoming a person.* Boston: Houghton Mifflin Company.

Weick, K. (1978, December). Educational organizations as loosely coupled systems. *Administrative Science Quarterly,* 23:541–552.

CHAPTER 2

Curriculum Planning

CURRICULUM is a work plan, a document that comes into being in order to influence the act of teaching. When it does so it becomes instruction (English, 1992, p. 17). Curriculum is both a view of the world and how it makes sense, and a "medium through which sense is made" (Page, 1993, p. 159). It is a kind of double looking glass. One looks in at the same time one is looking out.

Curriculum developers assume that creating a work document is purposive, i.e., it is done with the idea that a curriculum is a means to an end and not an end in itself. One constructs a curriculum to accomplish something. That something is what curriculum is constructed to attain. Part of the process of creating curriculum is to be clear about the intended consequences of curriculum. What has concerned many scholars is not so much the intended consequences of curriculum, but the unintended ones. The unintentional aims of curriculum have sometimes been called the "hidden curriculum" (Giroux, 1988, p. 51).

The second assumption is that curriculum is reproducible. It makes sense to not only create it, but print and distribute it. One can talk about the "third grade curriculum" or the "science curriculum" because it has characteristics of property. It is tangible. Curriculum, once constructed, can be stored, retrieved, put to use, debated, and subsequently altered as well as tested.

That curriculum is a means of reproduction is assumed to occur to either transmit the culture and the existing socio-political relations (sometimes referred to as hegemonic patterns or hegemonic curriculum) (Stanley, 1992, p. 97) or to transform and change them. The change away from hegemony to democratic socio-political relations is referred to as emancipatory practices (Stanley, 1992, pp. 193–196).

Most school people are unaware of the choices they make when considering developing curriculum. Some may believe that knowledge is neutral and that curriculum planning is simply an exercise in selection among a variety of possibilities. This is naive. Curriculum development is not a neutral activity, so planning its development cannot be neutral. Curriculum either reinforces the socioeconomic status quo or it attempts to change it. Not to decide ahead of time simply means allowing curriculum development to reinforce what already exists within schools and the relationships of schools to the larger socioeconomic order. Giroux (1983) calls this form of cultural domination the hegemonic-state reproduction model (pp. 222–225).

Knowledge exists within an hierarchical system of relations; that is, possession of knowledge benefits some and disadvantages others (see Foucault, 1980). Knowledge is not discovered. It is produced (English and Hill, 1994, p. 91). Knowledge should be considered relationally because it is neither stable nor inert. It is a kind of dynamic form of property. When knowledge is embraced by one culture as opposed to another that might be selected by a person not in that culture, it can be considered a form of cultural capital that privileges some in society and in the schools they attend (Giroux, 1983, p. 88; Darder, 1991, p. 12). The idea that schools advance students solely on the basis of achievement is a myth. Those in the most powerful positions in our society see to it that their children benefit the most from schooling by shaping the values in them to match their own socioeconomic positions and interests (Giroux, 1983, p. 164; McLaren, 1988, p. 163).

The curriculum planner must therefore consider developing curriculum chiefly as a political process and only tangentially a technical-rational process. If curriculum is a means of reproduction, exactly what is being reproduced? Who benefits from its reproduction? Much of the discourse surrounding the discussion of a "curriculum core" amounts to a contestation, a fight, if you will, between the various groups and the agencies that represent them, over whose curriculum will be taught in the schools. Public schools, as agencies of the state, reflect the dominant view of a culture that the poor in a society experience a curriculum that subordinates their modes of expression, language skills, and culture and fosters the impression that they are "inferior" and largely responsible for their own plight on the fringes of the social order (Giroux, 1983, p. 226). In this complex process within schools, an apparatus is constructed in which "knowledge functions as a reproductive force that partially serves to locate subjects within the specified boundaries of class, gender, and race" (Giroux, 1983, p. 224).

THE MATTER OF MORALS

Values are the clusters of choices within an overall discourse of ethics. A discourse is a conversation that can be verbal or written. When one considers what ought to go on in schools, choices are made that are guided by predispositions to believe or not believe in certain perspectives involving the meaning of life, the proper form of living, and proper action. These, in turn, rest on principles of good and bad, or morals. The principles of conduct are called ethics. Curriculum construction and planning occur within a discourse involving ethics.

That social institutions such as schools can and ought to be subject to ethical appraisals is a position embraced by the religious right in their attacks on public schools as amoral and godless places (see Manatt, 1995) to the more contemplative criticism of Mahatma Gandhi (Ayer, 1973, p. 27) who viewed material progress and moral progress as contradictory. Gandhi insisted that religion and politics could never be separated. "I do not know of any religion apart from activity. It provides a moral basis to all other activities without which life would be a maze of sound and fury signifying nothing" (Ayer, 1973, p. 41). Gandhi's view of religion, however, was not rooted in specific beliefs (sectarianism) but in religious values that there was a moral government of the universe (Ayer, 1973, p. 42).

Too often, curriculum planners surf through the thorny thickets of ethics naively assuming that a thorough exploration of ethical issues is not the stuff for school subjects when it can be shown that such subjects themselves represent overt and covert values, laden with moral choices. In the words of Kincheloe, McLaren, and Steinberg (1994):

> The technobabble of many educational leaders conveniently ignores questions of ethics. Is it ethical in a democratic society to educate students for stupidification? Is it ethical to separate the struggle for a more just and democratic society from the reform of education? It is a Disneyesque theory that states that all we have to do is fill students' heads with particular bits of knowledge and presto—the poverty and violence of their everyday lives disappear. (p. xv)

Henry Giroux (1992) views schooling as an instrumentalized "white, upper-middle-class logic that not only modulizes but actually silences subordinate voices . . . schooling is about somebody's story, somebody's history, somebody's set of memories, a particular set of experiences" (p. 14).

The curriculum development worker and planner should know that schools are contested battlegrounds in what critics of the left and right refer to as the culture wars (Short, 1986). Since any curriculum will embrace some values and not others, curriculum development is not a kind of engineering activity resting on "objective grounds" above the fight. To develop school curriculum means jumping into the arena and becoming a combatant, a contestant in determining whose values and whose ethics will be taught and learned in school classrooms. In the life of a young nation, it is perhaps one of the most critical of all discourses (see Gutierrez and Larson, 1994).

WHERE TO START THE PLANNING PROCESS

Thinking about developing curriculum that adheres to some sort of logical planning process that follows an organizational cycle of budgeting for resources requires that thoughts about it become public. That means creating a discourse and a forum for both the process and the content of the plan to be created.

Curriculum planning that involves the discussion of values, ethics and moral principles is much like engaging in a policy debate. To initiate the process requires honesty, patience, and respect for the points of view that will be expressed. The following steps are offered as a kind of policy discussion format and process that can be used to provide a framework for curriculum planning.

Setting the Ground Rules

Prior to any solicitation of positions of policy, a set of ground rules should be developed that establish the framework for the discourse about curriculum decision making. We offer these rules as important.

1. Respect for difference: Since a curriculum discussion is fundamentally a political and cultural set of choices, differences of points of view or perspective can be presented without derision or ridicule, providing that the presenters are open to debate and challenge of their views from those with a different opinion. No view is above challenge. No one perspective is considered sacrosanct.
2. Respect for minority opinion and the right to disagree: A vote does not guarantee the truth is embraced. Participants in the curriculum debate agree to work toward a consensus in which minority views are thoughtfully considered.
3. Separating challenge from attack, avoiding the language of "final vocabularies": Challenging a perspective, questioning its base or assumptions, is part of the discourse of debate. Attacking a position which involves fabrication, using unsubstantiated "facts," offering theories without data or bogus data, is eschewed. That these tactics have been those embraced by the religious right's attacks on public education have been documented by Mannatt (1995, pp. 136–142). Religious bigotry, or bigotry in any form, cannot be part of an informed discourse on curriculum planning. The use of theatrics and charges of New Age Satanism and witchcraft will doom any reasonable debate into a clash of political muscle rather than intelligent discussion. Bigotry leads to "moral ugliness and cruelty" (Tivnan, 1995, p. 249). What is needed for a discourse involving curriculum planning is to recognize that "we are all bundles of opinions and beliefs, of theories and prejudices about how we and our world are or

ought to be. To be able to make this imaginary leap is to have a well-developed moral imagination" (Tivnan, 1995, p. 250). Curriculum planning must transcend "local prejudice" and avoid what Rorty (1989) has called "the language of final vocabularies" (p. 91).

Developing a Vision of a Desired Future

The development of a future desired set of conditions as a kind of generalized "ideal vision" is the next step in the process. Visions can be societal, school system, or school-centered. Kaufman (1992) has referred to these as mega, macro, and micro in scope.

The more comfortable the participants are with the stated vision, the more likely it is that it will not disturb the fundamental aspects of the status quo. The less radical visions are easily acceptable to the participants. These are also most likely to represent people currently served reasonably well by the society, system, or school. The voices of those not served well may be unrepresented and silent.

Translating the Vision into Operational Indicators

Vision statements are nearly useless unless and until they are translated into things that can be seen, touched, or experienced in some way. If the vision statement indicates that "we believe that the mission of school is to challenge inequity in the larger society," what does this mean? How will we know if the school is really doing this? Here are some possible indicators for learning objectives:

- a refusal to consider anybody's history as the only one possible or desirable to be told
- learning to question self-evident truths
- learning to ask "who benefits from my believing this?"

Curricula is developed that

- include many perspectives
- avoid stereotyping people or their customs

- reject legitimacy as being established by power relationships that extend the privileges of the few, especially those currently in power

Suppose that the developers of curriculum desired to alter the relationships between teachers and students in schools, believing that authoritarian forms of power are represented and legitimized in schools. Curriculum would be constructed from a base where knowledge and power would be seen for the undemocratic relationships that are preserved in schools in the forms of pupil tracking and so-called "ability grouping" (Giroux, 1988, p. 195; Page and Page, 1993; Gutierrez and Larson, 1994).

It ought to be clear that curriculum development cannot be separated from school practices. Curriculum that does not challenge existing practices reinforces them. There are no "neutral" schooling practices. All are based on principles of perceived, stated and unstated, codes of moral behavior. Curriculum development is a form of ideology (Apple, 1979), a story with a closed ending established in the conditions of belief. Curriculum and schooling practices reinforce such beliefs (Apple, 1986).

Selecting Curriculum Content

The "stuff" that is picked to become part of the school's work plan is called the "content" of the curriculum. It can be represented as facts, processes, activities, experiences, skills and/or attitudes. Content is usually compatible with the vision. Curriculum content fits somebody's voice or opinion. It is an act of political choice rather than an "objective" search for truth (see Giroux and McLaren, 1992).

Developing a Curriculum Format

The format is simply the way the curriculum as a work plan has been packaged for teacher use in classrooms. Curriculum that excludes much teacher judgment has been criticized as becoming "teacher proof," which is a near impossibility. On the other hand, some curriculum is seen as an organizational device to preempt teacher judgment or decisions about delivering it. Apple (1986) has referred to this idea as "de-skilling" teachers (p. 32). The use of "technical control procedures," such as systems management and the application of behavioral objectives, serves to work against teachers advancing within a class-based economic system by stripping them of political power to shape their own work. This view stems from classical Marxian theory about those who do the work and those who own the enterprise. The de-skilling of teachers is viewed as a way to keep ownership from teachers of the work they perform.

The selection of a curriculum format should work against this trend by: (1) making the work to be done clear and (2) leaving "space" for the teacher to determine the best way to perform the work. The critics of this position would point out that to enable the "system" to determine the "what" of the classroom will most likely reinforce the division of labor already in existence in schools. This may be so, especially if the curriculum is falsely perceived to be a kind of "neutral" process of selecting "factual knowledge." However, curriculum development and planning are not conceptualized in this manner. They are, rather, products of a lively debate within a discourse emphasizing contestation and argument.

As the number of voices is expanded within the process of creating curriculum in schools, especially to include the previously marginalized or silenced members of groups not previously afforded any participation in the process, the selection of content will become more democratic, less hegemonical. Teachers cannot be the only voice to be heard in the process of curriculum development and planning. The views of parents and even students have a place in the discourse. Yet the fact of the matter is that curriculum is still the essential plan of work for teachers in schools. Differences between teachers are important to recognize as a way of individualizing teacher interest and competence, as well as utilizing the enormous well of energy teachers possess in delivering a curriculum that has personal meaning for them.

Formats should be "user-friendly," that is, make the teacher's job easier without deskilling what teachers do in classrooms. Huge "phone book" curriculum guides are antithetical to practical applications in classrooms. Being "user-friendly" reduces the size of guides and their intrusiveness, without leaving the faculty or the school helpless to include areas of the curriculum to be tested.

Building in Test Alignment

Curriculum alignment is the practice of making sure those areas of the curriculum tested are taught. Only in this established relationship can teaching be shown to have any influence in improving pupil performance as tested.

However, it should be pointed out that testing is perhaps the most blatant form of political control of the curriculum by agencies of the state. Testing is definitely contested political domain in American education. The power to test is the power to punish and reward students and whole classes of future citizens. That schools reward upper-class students and punish lower-class students via testing has been well documented (see LaBelle and Ward, 1994, p. 75). Alignment is one practice that neutralizes or depresses the predictability of the class structure to solely determine school success as defined by state norm referenced or even criterion referenced assessments.

Piloting the Curriculum Prior to Adoption

Curriculum should be piloted, that is, field tested, prior to widespread adoption. Too often curriculum is simply developed and sent out to teachers to utilize without seeing if the purposes for which it was created are actually met. Curriculum should be tested in live situations. Teachers should be asked to indicate what worked and what didn't. Feedback should be gathered and adjustments made where warranted.

Formal Adoption of the Curriculum

Because curriculum is as much a political process as a technical-rational one, boards of education should formally adopt curriculum. Many boards are reluctant to do so, feeling that their expertise is shallow, or that if the curriculum runs amuck on the fringe mores of a community, they can always claim it was not "theirs." Boards should formally adopt curriculum after they see the evaluative data by teachers, students, and even parents of new curricula.

Selling the Curriculum to the Community(ies)

New curricula should be controversial if it has anything new to say about anything. Lack of familiarity with new curricula will arouse skepticism and suspicion from some quarters in the various publics being served by their schools. New curricula should be piloted prior to adoption and explained in some detail, and the public should be invited to review the pilot results prior to adoption by the board. The public sometimes finds out about new curriculum by the grapevine or when their children flounder in it.

In one western school system, parents discovered that the language curriculum was "whole language" rather than traditional language by accident. As they began to find out how teachers in the schools were interpreting whole language they began to petition the board to change the curriculum. Many were resentful that prior to buying a home in the new school's area, they were not informed by anyone about the "innovations" that had been placed into the school. The lack of candor among educators about curriculum change led to parental suspicions that something was "afoot" that was trying to be hidden from them.

Evaluating the Curriculum

Formal evaluations of school curricula in local districts are very hard to find. If changing the curriculum is something that is "good" just by changing it, then new curricula is its own reason to change. In this scenario, the change itself is all there is to evaluate. This confusion of means and ends confounds the entire idea of evaluation and the assumption that curriculum is changed to alter the course of learning rather than an excuse for whatever learning might follow its adoption.

CURRICULUM AS THE MEANS FOR NATION BUILDING

It has been taken for granted that school curricula is intimately tied to the development of a nation and the creation of the means to change the lives of children. Arnove (1994) studied the impact of radical educational change in Nicaragua in the time period 1979–1993 as the schools were first taken over by the revolutionary Sandanistas and then later by a coalition government, which threw out the new curricula and textbooks of the revolutionaries. He offers a perspective that finds many echoes in the United States about school curricula and the purposes of education.

> It is unreasonable to expect an education system to play a major role, in the short-run, in transforming political cultures, bridging social and political divisions, and stimulating economic growth. In the long run, schools do affect the attainment of such goals, but more reasonable objectives might be that a country provide a basic education to all school age children that will equip them to be literate and numerate and have the knowledge and skills to acquire and interpret information about the basic social forces that affect their daily lives. (p. 205)

Schools are important means to national ends. As some nations try to change their international educational standing, they look to their schools to transform and to improve their international competitiveness. Two economists (Mulligan and Sali-i-Martin, 1995) proposed working out the wage rate of a zero-skilled worker by calculating an index of the returns a country receives from providing more education. The calculation is the actual earnings by the earnings of a zero-skilled worker. This wage-based approach is an index of human capital.

Yet how such "competitiveness" is to be defined and how minorities in various cultures are treated as a nation struggles to become "competitive" is great cause for concern. Just as in America, native peoples and their languages and cultures have been pushed to the social margins and relegated to positions of inferiority when compared to European models (see Fuentes and Carr, 1993; Gonzalez, 1994).

White Americans have been fond of quoting the idea of "the melting pot," which translated into crushing and erasing differences among mainly white immigrant groups. African Americans were never to be assimilated while Native Americans were shoved west onto reservations (Garcia, 1982, p. 39). In this idea, the ruthless suppression of any cultural differences or diversity was embraced as "the American creed." The idea that pluralism and the celebration of difference was a better goal was not adopted by the nation's schools as "healthy" until late in the 20th century. Even now it is controversial.

Even in societies that are relatively homogeneous, such as Japan's, the schools act as a mediating and sorting mechanism to maintain a rigid social class structure (see LeTendre, 1994). Calls for "global competitiveness" can become propellants for schools to perform even more overt forms of social class maintenance.

Curriculum acts as a kind of "cultural code," a grammar deeply rooted in socioeconomic class structure. As such, it functions as a sort of pedagogic habitués, a kind of living arrangement accompanied with specific pedagogic practices in schools (see Bernstein, 1990, p. 3). These practices serve as the regulators of human potential and reinforce the dominant cultural hegemony in schools that continue to marginalize alternative cultural concepts, languages, and modes of living and thought. Curriculum planning and development that solely concentrate upon the creation of a work plan that ignores the larger socioeconomic contexts in which schools function, merely reinforce the division of the haves and have-nots in every nation in which schools reproduce their society in the formal and hidden curricula in them.

Creating more effective schools by changing the curriculum first means recognizing the interests that are served by both the form and content of various kinds of curricula. For schools to become more successful with a wider band of pupils from all walks of life will mean adopting a posture of pluralism rather than erasure of difference. Monocultural curriculum cores that ignore the different stories that abound within various nations will continue to serve only the interests of the elite and the status quo in which their power is nested. As America becomes more diverse in the next century, cultural pluralism offers a way to enjoy difference within its borders. But it will clearly put an end to the wellspring of cultural control vested in the public schools as it has come to be defined so far. The battle for the curriculum and its allegiance to vested interests is high stakes contestation in which social class mobility is becoming more and more restricted to those earmarked as "culturally different."

REFERENCES

Apple, M. W. (1979). *Ideology and curriculum*. Boston: Routledge & Kegan Paul.
Apple, M. W. (1986). *Teachers and texts*. New York: Routledge & Kegan Paul.
Arnove, R. F. (1994). *Education as contested terrain: Nicaragua, 1979–1993*. Boulder, CO: Westview Press.
Ayer, R. N. (1973). *The moral and political thought of Mahatma Gandhi*. New York: Oxford University Press.
Bernstein, B. (1990). *The structuring of pedagogic discourse*. London: Routledge.
Darder, A. (1991). *Culture and power in the classroom*. Westport, CT: Bergin & Garvey.
English, F. W. (1992). *Deciding what to teach and test*. Newbury Park, CA: Corwin Press, Inc.
English, F. W. and Hill, J. C. (1994). *Total quality education: Transforming schools into learning place*. Thousand Oaks, CA: Corwin Press.
Fuentes, B. O. and Carr, S. E. (1993, January). Educational reform in Mexico. *International Journal of Educational Reform*, 2(1):12–18.
Foucault, M. (1980). In C. Gordon (Ed.). *Power/knowledge*. New York: Pantheon Books.
Garcia, R. L. (1982). *Teaching in a pluralistic society*. New York: Harper and Row Publishers.
Giroux, H. (1983). *Theory and resistance in education*. South Hadley, MA: Bergin & Garvey.
Giroux, H. (1988). *Teachers as intellectuals*. Westport, CT: Bergin & Garvey.
Giroux, H. (1992). *Border crossings*. New York: Routledge.
Giroux, H. and McLaren, P. (1992, April). America 2000 and the politics of erasure: Democracy and cultural difference under siege. *International Journal of Educational Reform*, 1(2):99–110.
Gonzalez, R. D. (1994, October). Race and the politics of educational failure: A

plan for advocacy and reform. *International Journal of Educational Reform,* 3(4):427–436.

Gutierrez, K. and Larson, J. (1994, January). Language borders: recitation as hegemonic discourse. *International Journal of Educational Reform,* 3(1):22–36.

Kaufman, R. (1992). *Mapping educational success.* Newbury Park, CA: Corwin Press.

Kincheloe, J. L., McLaren, P. and Steinberg, S. (1994). In Macdeo, D. (Ed.). *Literacies of power: What Americans are not allowed to know.* Boulder, CO: Westview Press.

LaBelle, T. J. and Ward, C. R. (1994). *Multiculturalism and education.* Albany, New York: SUNY Press.

LeTendre, G. K. (1994, April). Distribution tables and private tests: The failure of middle school reform in Japan. *International Journal of Educational Reform,* 3(2):126–136.

Manatt, R. P. (1995). *When right is wrong: Fundamentalists and public schools.* Lancaster, PA: Technomic Publishing Co., Inc.

McLaren, P. (1988). *Life in schools: An introduction to critical pedagogy in the foundations of education.* New York: Longman.

Mulligan, C. B. and Sali-i-Martin, X. (1995, March). A labour-income-based measure of the value of human capital. An application to the status of the United States as cited in Putting a value on people. In *The Economist.* (1995, June 24), p. 69.

Page, J. A. and Page, F. M. Jr. (1993, October). Tracking and research-based decision making: a Georgia school system's dilemma. *International Journal of Educational Reform,* 2(4):407–417.

Page, R. (1993). For teachers: Some sketches of curriculum. In Phelan, P. and Davidson, A. L. (Eds.). *Renegotiating cultural diversity in American schools.* New York: Teachers College Press, pp. 159–194.

Rorty, R. (1989). *Contingency, irony, and solidarity.* Cambridge: Cambridge University Press.

Short, I. (1986). *Culture wars.* Boston, MA: Routledge and Kegan Paul.

Stanley, W. B. (1992). *Curriculum for Utopia.* Buffalo, New York: SUNY Press.

Tivnan, E. (1995). *The moral imagination.* New York: Simon and Schuster.

CHAPTER 3

World-Class Curriculum

STUDENTS graduating from American public schools today enter a world dominated by interconnecting, global linkages; a world requiring successful adults to be able to function effectively in a global community. Given the plethora of international assessment results comparing this country with others, the average high school graduate in the United States today does not have these necessary skills. While the most outstanding high school graduates in the United States appear to have skills comparable with the most outstanding high school graduates in other industrialized nations, there are simply fewer of these outstanding students in the United States (Gandal, 1994). The percentage of the age eighteen cohort who takes and passes at least one advanced, subject-specific exam is significantly higher in England and Wales, France, Germany, and Japan. In the United States, 7 percent of the age cohort take Advanced Placement Exams and 4 percent achieve a grade of three or above. In England and Wales, 31 percent take at least one exam and 25 percent pass; in France 43 percent take and 32 percent pass; in Germany 37 percent take and 36 percent pass, and in Japan 43 percent take and 36 percent pass (p. 96). In all of these countries, except the United States, college-bound students are required to pass content-based tests as part of the admission process. The United States has no such requirement.

The skill level of high school graduates is even lower. Joseph Gorman, chairman of the Business Roundtable Task Force, recently stated, "The harsh reality is that generally our schools are not teaching what is required for success in our increasingly complex global economy. We give them diplomas rather than knowledge and skills that would help them land a good job" (Gorman, 1995). In 1990, the Business Roundtable committed itself to a ten-year effort to improve the quality of public education, state by state. While claiming steady progress in this initiative, Gorman views the improvement as too slow. In June 1994, the Roundtable established a new agenda composed of seven components:

- standards that set high academic expectations linked to a system of school-based sanctions and rewards documenting accountability
- school autonomy that provides the resources and decision-making flexibility to hold schools accountable for student learning
- professional development leading to continuous improvement for teachers and administrators
- parent involvement that supports choice and active involvement in the creation of the learning environment
- learning readiness that focuses on pre-school skill development for those children more prone to school failure
- technology to access knowledge
- safety and discipline in the learning environment

The Roundtable has taken the position that standard setting is the most essential of the nine components. Often criticized for short-term support for education, this time business leaders are planning for a long-term sustained effort that includes collaborative efforts with state education agencies, public information campaigns, and a sustained political presence. "We do not think of education as just a high priority. We see it as nothing less than an issue of economic, political, social, and cultural survival" (Gorman, 1995, p. 51).

ROLE OF NATIONAL STANDARDS

In most industrialized countries, the national government plays some role in establishing and coordinating rigorous, publicly-known standards. The establishment of national content, curriculum standards in the United States has become a raging debate. Some have referred to it as the "hottest item in education reform today" (Lewis, 1995, p. 745).

Since 1989 the federal government has funded subject-area groups and associations to develop content and performance standards in areas related to the national education goals. These include science, history, civics, language arts, geography, the arts and foreign language. In addition, private and public funding has supported national efforts to establish standards in physical education, health, and social studies. Financial support for these efforts has exceeded $10 million. Each of these standard-setting initiatives includes a broad array of participants that represent teachers, scholars, administrators, parents, and others.

While the idea of establishing national standards has been present for a long time, the current initiative got underway at the first National Education Summit held in Charlottesville, Virginia, in 1989. At that meeting President Bush and the National Governors Association, led by Arkansas governor, Bill Clinton, announced six national goals. In 1994 two additional goals were added. The present national goals are:

1. All children in America will start school ready to learn.
2. The high school graduation rate will increase to at least 90 percent.
3. All students will leave grades 4, 8, and 12 having demonstrated competency over challenging subject matter including English, mathematics, science, foreign language, civics and government, arts, history and geography, and every school in America will ensure that all students learn to use their minds well, so they may be prepared for responsible citizenship, further learning, and productive employment in our nation's modern economy.
4. The nation's teaching force will have access to programs for continued improvement of their professional skills and the opportunity to acquire the knowledge and skills needed to instruct and prepare all American students for the next century.

5. United States students will be first in the world in mathematics and science achievement.
6. Every adult American will be literate and will possess the knowledge and skills necessary to compete in a global economy and exercise the rights and responsibilities of citizenship.
7. Every school in the United States will be free of drugs, violence, and the unauthorized presence of firearms and alcohol and will offer a disciplined environment conducive to learning.
8. Every school will promote partnerships that will increase parental involvement and participation in promoting the social, emotional, and academic growth of children.

In July of 1990, the National Education Goals Panel was established to assess and annually report state and national progress toward achieving those goals. In 1992, the chair of the National Education Goals Panel, Governor E. Benjamin Nelson, stated, "The National Education Goals are an integral part of the challenge to make quality and competitiveness hallmarks of America once again. We, on the National Education Goals Panel, are committed to providing leaders at all levels with a clearer vision of what needs to be improved to transform our schools and to provide world-class, high-performance learning" (Lancaster and Lawrence, 1992–93, p. iii).

The 1994 report (National Education Goals Report), *Building a Nation of Learners,* marked the fifth anniversary of the summit. In March 1994, Congress adopted and the president enacted the *Goals 2000: Educate America Act,* which expanded the role of the Goals panel. In addition to reporting the progress made by the nation and the states relative to the national goals, the panel's responsibilities include (p. 3)

- building a national consensus for education improvement
- accelerating progress by reporting on promising or effective actions being taken at the national, state, and local levels to achieve the Goals
- identifying actions that federal, state, and local governments should take to enhance progress toward

achieving the Goals and to provide all students with a fair opportunity to learn
- working in partnership with the newly created National Education Standards and Improvement Council (NESIC) to review the criteria for voluntary content, performance, and opportunity-to-learn standards reflecting high expectations for all students

The eighteen-member Goals Panel is made up of eight governors, four members of Congress, four state legislators, the U.S. Secretary of Education, and the assistant to the President for Domestic Policy.

The 1994 report is organized around sixteen core indicators for achieving the eight goals. The panel believes that the heart of the goals process is informed decision making. Communities are encouraged to develop their own data-based decision making process that will enable them to (National Education Goals Report, 1994, p. 16):

- adopt and adapt the National Education Goals to reflect high expectation for all learners and cover a lifetime of learning, from the preschool years through adulthood
- assess current strengths and weaknesses, and build a strong accountability system to measure and report progress regularly toward all of the goals
- set performance milestones to serve as checkpoints along the way

The core indicators of Goal achievement are as follows:

1. Children's health index: Reduce the percentage of children born with two or more health or developmental risks.
2. Immunizations: Increase the percentage of two-year-olds who have been fully immunized.
3. Family–child reading and storytelling: Increase the percentage of three- to five-year-olds whose parents read or tell stories to them on a regular basis.
4. Preschool participation: Reduce the gap in preschool participation from high- and low-income families.
5. High school completion: Increase the percentage of nineteen- to twenty-year-olds with a high school diploma.

6. Mathematics achievement: Increase the percentage of students who meet Goals Panel standards as measured by the National Assessment of Educational Progress (NAEP).
7. Reading achievement: Increase percentage of students who meet Goals Panel standards as measured by the NAEP. *Note:* During the next three years the Panel expects to be able to add history, geography, science, and arts achievement to the list of core indicators.
8. International mathematics achievement: Improve standing internationally.
9. International science achievement: Improve standing internationally.
10. Adult literacy: Increase percentage of adults who score at or above Level 3 in prose literacy.
11. Participation in adult education: Reduce gap of adult education participation between adults with high-school diplomas and those without high school diplomas.
12. Participation in higher education: Reduce gap between black and white high school graduates who enter college and complete a college degree. Reduce the gap between white and Hispanic high school graduates who enter college and complete a college degree.
13. Overall student drug and alcohol use: Reduce the number of tenth graders who report the use of illicit drugs and alcohol.
14. Sale of drugs at school: Reduce the number of tenth graders reporting someone offered to sell them illegal drugs during the previous year.
15. Student and teacher victimization: Reduce percentage of threatened or injured persons.
16. Disruptions in class by students: Reduce percentage of reported disruptions.

No core indicators have yet been established for Goal 4: Teacher Education and Professional Development and Goal 8: Parental Participation. However, it is the intent of the Goals Panel to establish indicators for these two goals in the future.

The work of the committees developing standards in specific

content areas is directly related to the national education goals and will be used by the creators of the National Assessment of Educational Progress to design the national assessments at grades 4, 8, and 12 and track progress in meeting the national goals. This relationship of the current standards setting movement must be placed into the context of how these standards will be used to set a national curriculum and measure national achievement.

The present controversy surrounding the standards-setting initiative must be viewed within the context of how and why the work was organized and funded and the intent of the governors at that important 1989 Education Summit to establish a set of high expectations for what students graduating American high schools should know and be able to do.

The national standards-setting movement is a means to an end. The end is to dramatically improve the quality of the public educational system through the establishment of high expectations for all students so that the United States can retain its position as a world power. It is an economic goal supported by the elite of the business community to maintain an economic edge. It is motivated by profit. If other industrialized countries have more workers with the intellectual and educational capability to create the products and services needed in the 21st century, then other industrialized countries will become known for their "cutting edge" expertise and will, over time, cause a shift in the world power base. It is, at its base, an economic initiative with an economic goal.

At the same time, there are powerful forces within this country that want to maintain the economic stratification that presently exists. Students who succeed within the present system of education do so because the educational environment of the home supplements the educational environment of the school. Children from families with higher levels of socio-economic status (SES) consistently do better on national achievement tests than children from families with the lowest SES where the curriculum of the home does not support the curriculum of the school. Groups supporting the continuation of the public educational system as it exists today are generally those who have benefited from the present system. Children from families who have been victims of the deficiencies of the present system generally have no political voice and they are unable to find a way for their voice to be heard. This author would argue that the problem is not with the children or the parents of the children. The problem is with the system. The variations in student achievement and the relationship of that achievement to SES is an expected outcome of the present system of public education. Trying harder won't fix the problem. Development of a state-wide system of sanctions and rewards won't solve the problem. Establishing eight National Goals and identifying indicators for tracking progress in achieving those goals won't solve the problem. Children from higher SES levels will continue to do better than children from lower SES levels. Schools that have been successful in changing the predicted outcomes have been schools that have changed how the system works. These schools have not only established clear standards for what students should know and be able to do, these schools have changed the delivery system for how to achieve the standards.

The present standards movement was created as a means for developing a national curriculum. The benefit of the movement may yet be reaped at the local school level as educators, parents, and community members engage in dialogue over the American Dream—who has access to it and who does not. The American Dream goes something like this: if you work hard, you can accomplish whatever you want. The implication seems to be that if you don't accomplish want you want, then it was your own fault and you simply didn't work hard enough.

Most parents, school board members and the community at large believe that educators know how to educate all children. In reality, most educators do not know how to educate all children at high levels nor do they believe that all children are capable of learning at high levels. A recent survey of teachers in the state of Kentucky revealed that only 34.7 percent of the teachers surveyed believed that all children could learn at high levels (*Statewide Education Reform Survey,* 1994). This is a significant finding in a state that passed the largest tax increase the state had ever known to support the 1990 Kentucky reform (Steffy, 1993). At the time the Kentucky reform was first enacted, educators were asked to estimate how much money it would take to solve the educational problems plaguing the state and that money was provided. Still, a third of the teachers in Kentucky classrooms do not believe in one of the fundamental principles upon which the reform was built. The reality of the situation is that, as an educational profession, we have never achieved this expectation. Yet we have never been honest with the public and said, "I can't teach this child; I don't know how within the present system of education!" We have continued to take the money generated by the student's attendance in school until the student decides to drop out and then we blamed the student or the parents or society for the failure. Most kindergarten and first grade teachers can identify those students who are going to fail within the present system. To change the future for these children means changing the system. Few educators are prepared to attempt the process. So the system grinds on. Some students do exceedingly well. Estimates vary regarding the percentage of students who attain world-class standards. Most estimates fall between 15 and 20 percent of those students presently entering kindergarten. Many students will fail, drop out, or graduate as functional illiterates. Estimates of the percentage of children in this class range from 15–30 percent depending on what state or district you happen to be in. It is not unusual for an urban high school to have a drop-out rate in excess of 60 percent of the age cohort. These districts tend to serve large numbers of children from low socioeconomic classes. On the other hand, suburban districts in high socioeconomic areas like to brag about their low drop-out rates.

Changing the system means expanding the potential pool of educated young adults capable of becoming doctors, lawyers, merchants, and CEO's. Just how many of these do we need? The present system serves as a sorting mechanism that enables the children of more affluent families to achieve a higher status and gain access to the best opportunities to succeed. Politicians are often heard talking about the establishment of a "level playing field" where all children have an equal opportunity to succeed. Even the indicators for achieving the National Goals refer to decreasing the achievement spread between the black and Hispanic population and the white student population. It sounds good. It buys votes. But, when it comes to changing the system so that it becomes a reality, the power base of the beneficiaries of our present system rally and become politically active.

The national standards movement has become a lightning rod for national debate about who should have access to the benefits of public education. The establishment of standards alone won't solve the issues involved.

THE FIGHT FOR ACADEMIC INSTRUCTIONAL TIME

Schools today are organized just about the way they were ten or twenty or fifty years ago. The school year and school day have not changed significantly. The Carnegie Unit is still used as the input measure for high school graduation. Assemble enough Carnegie Units based on a grade of D or above and you will graduate high school. What the graduating high school senior knows and is able to do is of less importance than the number of Carnegie Units achieved. The United States is the only nation in the world that uses the Carnegie Unit as the ultimate indicator for high school graduation. From our perspective, it is one of the flaws in our public education system. The standards movement would disrupt this measure of seat time as an indicator of academic achievement.

STANDARD SETTING AT THE STATE LEVEL

The national standards movement is just one example of the forces attempting to establish a new mechanism for determining what students know and are able to do. All of the fifty states are involved in some form of standards setting at the state level. Many of these state initiatives call the formation of state standards by a different name. While they are referred to in a number of ways, many states are establishing what they call curriculum frameworks or content standards and using these standards as the basis for the development of the state assessment system ("Struggling for Standards," 1995, pp. 23–35).

These efforts are seen as a mechanism for promoting school reform and are being developed utilizing a set of six guiding principles (Curry and Temple, 1992, pp. 21–23):

- balance—developed around a balanced core of common learning
- assessment—focus on results
- interdisciplinary learning—promote interdisciplinary learning
- active learning—encourage active learning
- diversity—recognize and respect student diversity
- thinking curriculum—develop thinking skills

Table 3.1 outlines what the standards are called, whether they are voluntary or mandated, and how they relate to the state assessment system for each state.

The standard-setting efforts at the state levels were organized as a unified effort within the state. Generally, these standards were not developed by independent content associations using different formats, terminology, and procedures. However, like the national standards they do tend to be stated in rather general terms. Designing classroom instruction utilizing these state standards requires additional refinement at the district level to enable the district to decide exactly what the standards mean for instruction at a specific grade level. Like the national standard-setting movement, state frameworks and content standard have used the pioneering work of the National Council of Teachers of Mathematics as a guide. The objective for this movement at both the state and national level is to enable all schools to offer a challenging curriculum which can be tied to district, state, and national assessments (Smith, Fuhrman, and O'Day, 1994).

In many respects the state standard-setting initiative has succeeded in accomplishing the objective of the national standard movement with a lot less controversy. True, states such as Pennsylvania, Virginia, and Colorado have been challenged by opponents of outcomes-based education and modifications in the state standards that have been made are a result of these challenges. Nevertheless, no one is suggesting that the state does not have the right and the power to establish these standards.

Table 3.1. State level standard setting initiatives.

State	Standard Name	Mandated or Voluntary and Relation to State Test
Alabama	Courses of study	Mandatory
Alaska	Student performance standards	Voluntary, state is designing assessment system
Arizona	Essential skills	Mandatory if the subject is mandatory. By the year 2000 graduates must demonstrate proficiency in essential skills
Arkansas	Curriculum frameworks	Voluntary but state tests built around them
California	Curriculum frameworks	Voluntary but tied to state assessment
Colorado	Content standards	New state tests due in 1996–97 tied to content standards
Connecticut	Guides for curriculum development	Voluntary but tied to state assessment
Delaware	Content standards with performance indicators	Ultimately mandatory
Florida	Curriculum frameworks	Voluntary but tied to state assessment
Georgia	Quality core curriculum	Voluntary but currently tied to state assessment
Hawaii	Content standards and performance standards	Mandatory
Idaho	Curriculum frameworks and content standards	Voluntary
Illinois	Academic standards	Related to state assessment system
Indiana	Content standards and performance standards	Voluntary
Iowa	No state curriculum standards; districts must develop goals	If developed, voluntary
Kansas	Content standards	Related to state assessment system
Kentucky	Academic expectations	Mandatory
Louisiana	Content standards	Mandatory
Maine	Discipline standards	Undecided but standards will form the basis for state assessment system
Maryland	Learning outcomes	Mandatory
Massachusetts	Curriculum frameworks	Voluntary but tied to state assessment system
Michigan	Content standards	Tied to state assessment in core content areas
Minnesota	Basic requirements	Mandatory

Table 3.1. (continued).

State	Standard Name	Mandated or Voluntary and Relation to State Test
Mississippi	Curriculum structure	Voluntary but tied to state assessment system
Missouri	Curriculum frameworks	Voluntary, eventually will be tied to state assessment system
Montana	School accreditation standards	Program area standards are mandatory
Nebraska	Curriculum frameworks	Voluntary, no state assessment system
Nevada	Courses of study	Mandatory
New Hampshire	Curriculum frameworks	Voluntary but basis for state assessment system
New Jersey	Core curriculum standards	Undecided, future state assessment will be tied to standards
New Mexico	Competency frameworks	Mandatory
New York	Curriculum frameworks	Undecided, but will be tied to state assessment system
North Carolina	Standard course of study	Mandatory
North Dakota	Curriculum frameworks	Voluntary, not tied to state assessment system
Ohio	Curriculum frameworks	Voluntary but tied to state assessment system
Oklahoma	Priority academic student skills	Required, test under development
Oregon	Content standards	Mandatory
Pennsylvania	Student learning outcomes and model state content standards	Mandatory
Rhode Island	Curriculum frameworks	Voluntary, emerging test tied to standards
South Carolina	Curriculum frameworks	Voluntary, test revision underway tied to standards
South Dakota	Benchmarks or content standards	Voluntary, possibility of state tests to complement standards
Tennessee	Curriculum frameworks	Mandatory but allow a lot of freedom at the district level State testing program in revision
Texas	Essential elements	Districts responsible for testing essential elements. Essential elements tied to state's criterion referenced testing system

(continued)

Table 3.1. (continued).

State	Standard Name	Mandated or Voluntary and Relation to State Test
Utah	State core curriculum	Mandatory but assessment standards are voluntary
Vermont	Content standards, performance standards	Districts must adopt standards at least as rigorous as state standards. State assessment system under development
Virginia	Standards for learning	Under debate, not expected to be mandated. Will be tied to graduation requirements and state tests
Washington	Essential academic learnings	Mandatory by the year 2000. Standards tied to statewide testing
West Virginia	Program of study	Mandatory, tied to performance standards through graduation requirements
Wisconsin	Content standards, performance standards, opportunity to learn standards	Voluntary, relation to state assessment to be determined
Wyoming	Common core of knowledge, common core of skills	Districts must develop local standards and create own assessment system

Whether these efforts will bring school districts closer to the establishment of a world class curriculum is yet another question.

WHAT OTHER COUNTRIES DO

The fundamental belief that all children can learn at high levels does not appear to be the basis for instructional practices in other industrialized countries. This is not to say that educators in these countries do not believe in this principle. In most industrialized countries, all children are not expected to achieve the same standards. If a student wants to graduate with a certain type of diploma, the student knows he/she will have to pass certain exams. The standards are set and publicly known, and students are responsible for achieving or not achieving them. In Germany, for instance, public education is primarily the responsibility of the sixteen states (Gandal, 1994). The state's role is one of coordination. While compulsory education lasts for nine to ten years, students enter one of three tracks after completing primary school. The tracking process usually begins after the fourth grade and is primarily determined by how well students perform in primary school. Approximately one-third of the students in Germany attends the most basic level, which lasts until ninth grade, and another one-third attends the Gymnasium, which is considered the most rigorous. In order to enter the university, students must achieve a passing score on the Abitur (Gandal, 1994, pp. 46–48).

In Japan, where ability grouping is prohibited in grades 1–9, students must pass competitive tests to be admitted to high school. There are no single standardized tests for all students to pass in order to be admitted to high school. Each high school develops its own tests. Although high school is not mandated in Japan, nearly 95 percent of the students attend. While only a small percent of students attend vocational and technical high school, about one-quarter of the high school graduates pursue some form of vocational training. A large percentage of students apply for admission to the university. In 1990, this figure was reported to be 43 percent (Gandal, 1994, pp. 58–59). In France the highest high school degree is called the baccalaureate. It is awarded based on a student's performance on a set of exams. In 1950, only 5 percent of the age eighteen cohort was awarded a baccalaureate degree. By 1992, that figure had risen to 51 percent.

In the United States there are no nationally required tests for graduation from high school or admission to colleges and universities. While most colleges require students to submit scores for the ACT or the SAT for admission, these tests are not tied to any particular curriculum. Of the 3,600 colleges and universities in the United States, nearly one-third operate under an open admissions policy with no admission standards. Students in other countries know that if they want to attend college, they must pass rigorous, subject-matter tests. The standards are set, and if the students want access to the university, they meet the standard. On a recent trip to Scotland, I had an opportunity to visit two different homes. In both cases, the families had teenage children and in both cases the children were studying and appeared to take pride in sharing examples of their work with visitors from America. One of the visits took place on a beautiful, sunny (I was told this was a bit unusual for rainy Scotland) school holiday. The visit caused me to think about whether most high school students in the United States would spend the day studying. In both cases, the youngsters were studying for high school exit exams that would not take place for another year.

Are the exams in other countries more rigorous? This is a difficult question to answer. The American Federation of Teachers (AFT) has undertaken a study to determine the answer. While no definitive answer to this question has been determined at the present time, the characteristics of the most rigorous exams in the content area of biology is available. The countries include England and Wales, France, Japan, Germany, and the United States. The exams were evaluated in three areas: exam format, exam content, and other issues (Gandal, 1994, pp. 101–104).

Analysis of the exam format was divided into three areas: exam length, choice, and question type. Among the biology exams reviewed, the A-level exam in England and Wales was considered the most rigorous, both in length and in requiring students to "display a higher level of discipline and fortitude." This exam lasts for nine hours. The exams in France and England and Wales allow students some choice among questions; the exams in the United States, Germany and Japan do not. Question type differs significantly among the five countries with Japan and the United States as the only countries to use multiple choice questions. It was reported that 60 percent of the AP biology score in the United States is made up of responses to multiple choice questions. France, Germany, and England and Wales rely only on open-ended response items. Generally, open-ended response items require students to "make and defend judgments, demonstrate scientific method, explain complicated logic in clear prose, and otherwise show

how they arrived at their answers" (Gandal, 1994, p. 102). Multiple choice items do not typically assess these types of skills.

Exam content was reviewed for types of performances assessed, breadth versus depth, and complexity of knowledge. There are few questions on the European exams that require students to recall single bits of knowledge. Rather, these exams require students to innovate, show their work, explain their answers and back up their conclusions (Gandal, 1994, p. 102). While 40 percent of a student's score on the AP exam is derived from responses to open-ended questions, 60 percent of the score is based on responses to multiple choice questions. An answer to the depth and breadth question did not reveal a definitive answer. Some of the exams reviewed required students to respond in greater depth to a narrow body of knowledge. Others, such as the AP, required students to display a wider range of knowledge. The issue of complexity of knowledge was also difficult to assess due to the differing formats of the tests.

Other issues considered included grading standards, preparation, scope of the examination system, and how much each exam counts. The bottom line of the comparisons seemed to boil down to the fact that in all countries except the United States, successful passage of these tests was required for admission to college. With only 4 percent of students in the United States passing the Biology AP exam with a grade of 3 or above and the average among the other countries reviewed being somewhere between 25 percent and 32 percent, one has to wonder about the long-term impact on the scientific presence of the United States in the world community.

OPPORTUNITY TO LEARN

Other countries appear to be less interested in educating all students to the same level of achievement or assuring all students that there is an equal opportunity to learn (OTL). The phrase *opportunity to learn* is included in the *Goals 2000: Educate America Act*. It means "the criteria for, and the basis of, assessing the sufficiency or quality of the resources, practices, and conditions necessary at each level of the education system (schools, local educational agencies, and states) to provide all students with an opportunity to learn the material in voluntary national standards or state content standards" (*The National Education Goals Report,* 1994, p. H1626).

While the development of opportunity to learn standards is voluntary, any state that wished to receive Goals 2000 funding from the federal government had to develop opportunity to learn standards. For example, in requesting Goals 2000 funding, a state had to indicate how it would address OTL standards, how OTL standards related to content standards, and how OTL standards would be used (Porter, 1995, p. 21). In addition, $2 million was authorized for fiscal year 1994 as OTL development grants. These competitive grants were awarded to develop a set of voluntary OTL standards and a listing of model OTL programs. The role of OTL standards in the national standards movement is uncertain at this point. The debate is closely related to school equality and accountability issues. If students are denied access to certain benefits based on test scores attached to state standards, should school districts be held accountable for assuring that all students in the district had an adequate opportunity to learn the standards? Further, does the definition of opportunity to learn include just inputs, such as books and materials, or does it include assurance that the instructional strategies employed by teachers enabled all students to achieve at high levels?

The debate over opportunity to learn has resulted in discussion about a new type of curriculum alignment for the 21st century (Steffy, 1995). This type of alignment refers to the

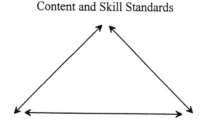

Figure 3.1. 21st century alignment.

relationship among content and skill standards, performance standards, and opportunity to learn standards. Content and skill standards

> spell out the subject-specific knowledge and skills that schools are expected to teach and students are expected to learn. Standards-setters have adopted the shorthand phraseology "what students should know and be able to do."
>
> Performance standards gage the degree to which students have met the content standards. They specify how students must demonstrate their knowledge and skills and at what level. They answer the question, "How good is good enough?"
>
> Opportunity-to-learn standards define the kinds of resources that students need to meet the content standards, including instructional materials, a safe environment, and well-trained teachers. ("Struggling for Standards," 1995, p. 8)

Figure 3.1 depicts this new type of curriculum alignment. The basis for this concept was discussed in Chapter 1 as the relationship of the written, taught and tested curriculum. Here, we have placed it in the context of the standards movement. The written curriculum is further defined as content and skill standards; the taught curriculum is further defined as opportunity to learn standards, and the tested curriculum is defined as performance standards.

The concepts of design and delivery, focus and connectivity, and context and content alignment all apply as described in Chapter 1.

WHAT THE GRADUATING SENIOR IS EXPECTED TO KNOW AND BE ABLE TO DO

Just what would a high school graduate who meets the standards identified in ten key content areas know and be able to do? Table 3.2 is an abbreviated version of the high school standards published in the April 12, 1995, *Education Week* supplement entitled, "Struggling for Standards." Additional groups are currently working to establish standards in areas such as bilingual education, vocational education and economic education. This summary was compiled to provide the reader with a

Table 3.2. What graduating seniors in the United States should know and be able to do—A condensed summary.

Content Area	Standard
Reading and language arts	*Note:* No integrated reading and language arts standards have been established. Federal funding for the project has been withdrawn. The reading group and the language arts group are now working separately.

Mathematics	1. Use mathematical procedures to solve real-world problems both within and outside mathematics
	2. Understand and use mathematical vocabulary and concepts to:
	    reflect about the clarity of thinking
	    formulate definitions
	    express ideas orally and in writing
	    read with understanding
	    ask clarifying and extending questions
	    appreciate mathematical notation
	3. Use mathematical reasoning skills to construct, judge and follow logical arguments, formulate counter examples, and make and test conjectures
	4. Recognize and use mathematical connections among mathematical topics and other disciplines
	5. Use algebraic concepts to represent situations, construct graphs and tables, and solve equations and inequalities
	6. Understand functions in order to analyze and represent relationships using graphs, verbal rules and equations
	7. Understand, interpret, and draw three-dimensional geometric figures in order to classify objects and deduce relationships
	8. Understand geometry from an algebraic point of view in order to deduce properties, identify congruent and similar figures, and analyze Euclidean transformations
	9. Apply trigonometry to problem situations and explore real-world phenomena using the sine and cosine functions
	10. Understand statistical concepts, such as sampling, central tendency, variability and correlation in order to design experiments and analyze the effects of data transformation
	11. Understand and use the concept of probability
	12. Represent problem situations using discrete structures in order to analyze algorithms, finite graphs using matrices, and solve enumeration and finite probability problems
	13. Understand the conceptual underpinnings of calculus
	14. Understand the underpinnings of a variety of mathematical structures

Science	1. Understand and conduct scientific inquiry 2. Understand physical science concepts such as the structure of atoms and matter, chemical force, motions and reactions. Understand the conservation of matter and the interaction of energy and matter

Table 3.2. (continued).

Content Area	Standard
	3. Understand life science concepts such as biological evolution, the cell, and the molecular basis of heredity. Understand the organization and interdependence of living systems. Know how the nervous system affects the behavior of organisms
	4. Understand the origin and evolution of the Earth and the universe including geochemical cycles and energy
	5. Understand the science of technology
	6. Develop an understanding of science in person and social perspectives, such as population growth, environmental quality, and technology at the local, national and global level
	7. Understand the history of scientific knowledge
Social studies	1. Analyze, predict, apply, compare, demonstrate, interpret, construct, and explain the impact of culture on human behavior
	2. Demonstrate, apply key concepts, identify and describe, employ processes, investigate, interpret, and analyze the impact of time continuity and change
	3. Refine mental maps; create, interpret, and use information; examine, interpret, and analyze physical and cultural patterns to gain an understanding of people, places and environments
	4. Understand individual development and identity
	5. Apply concepts such as role, status, and social class in order to understand, analyze and explain the behavior of individuals, groups, and institutions
	6. Understand the concepts of power, authority, and governance by examining, explaining, analyzing, comparing, and evaluating issues related to these concepts
	7. Analyze, explain and understand the concepts of production, distribution, and consumption
	8. Understand the role science and technology play in society
	9. Analyze, explain, and illustrate how language, conditions, and relationships operate to form global connections
	10. Analyze, evaluate, explain, and practice civic ideals, action and policy
U.S. and world history	*Note:* Controversy regarding these standards has led to the formation of two independent review panels. The panels, commissioned by the Council for Basic Education, will review the work for "scholarly merit, balance, and feasibility for practitioners." The standards, released in October 1994, have been criticized for overemphasizing multiculturalism and portraying the United States and the West as repressive. They were developed with the involvement of over 1000 educators representing more than 30 related associations at a cost of $2 million in federal funding. (Diegmueller, June 21, 1995, p. 8)

(continued)

Table 3.2. (continued).

Content Area	Standard
U.S. history	1. Understand stages in European colonization and Spanish conquest from 1492–1700
2. Understand the colonization and settlement patterns in America from 1585–1763 and their impact on family life, gender roles, women's rights, political rights, economic and labor issues, and slavery
3. Understand the causes, effects, and issues related to the American Revolution (1754–1820s)
4. Understand the issues and practices of government at the state and national level including the development of the constitution, political party system, and Supreme Court
5. Understand the impact of expansion and reform during the period of 1801–1861 as it relates to Native Americans, slavery and the reorganization of the political party system
6. Understand the causes and impact of the Civil War and the period following the war (1850–1877)
7. Understand the development of an industrialized United States during the period of 1870–1900 and the impact of this development on big business, urbanization, and politics
8. Understand the emergence of modern America from 1890–1930 as a world power
9. Understand the causes of the Great Depression and how it resulted in the New Deal
10. Understand the causes and effects of World War II
11. Understand the postwar era of 1945–1970 and how it affected economic, political, and social development
12. Understand major social, political and economic developments from 1968 to the present time
World history	1. Understand the biological and cultural beginnings of human society that resulted in the emergence of agricultural societies
2. Understand the early emergence of pastoral peoples (4000–1000 BCE) and the impact on political, social, and cultural issues
3. Understand the emergence of classical traditions, major religions, and giant empires (1000 BCE–300 CE)
4. Understand the causes and consequences of the expanding civilizations during the period of 300–1500 CE including communication, trade, and cultural exchange; the growth of towns and cities; the patterns of crisis; and the expansion of states on America
5. Understand the consequences of global expansion from 1450–1770
6. Understand the causes and consequences of political, agricultural, and industrial revolutions during the period of 1750–1914 including global trends and patterns of nationalism

Table 3.2. (continued).

Content Area	Standard
	7. Understand the causes and consequences of World War I, World War II, the Cold War, and peace-keeping efforts during the 20th century 8. Understand the promises and issues impacting the last half of the 20th century regarding population expansion, increasing economic interdependence, and human rights 9. Understand major world-wide trends that will impact the 21st century
Civics	1. Explain terms such as civic life, politics, and government and explain arguments supporting need for politics and government 2. Compare, contrast, and evaluate positions related to the importance of law, the need for a civil society, and the relationship between political and economic freedoms 3. Explain the nature and purpose of the Constitution and how it has shaped American society 4. Understand, explain, evaluate, and take and defend positions regarding a variety of forms of government 5. Explain and describe the concept of a political culture, the importance of shared political beliefs, and the impact of political conflict 6. Understand, explain, take and defend positions and evaluate the concept of liberal democracy as it relates to America 7. Understand, explain, take and defend position, and evaluate how power and responsibility are distributed, shared, and limited by the United States Constitution 8. Understand, explain, take and defend positions, and evaluate issues related to judicial protection of individual rights, the relationship of state and local governments, providing money for social services, the development of foreign and national policy, and the formation and function of national institutions 9. Evaluate, take and defend positions related to setting a public agenda, the role of public opinion in American politics, the influence of public media, role of political parties and election campaigns, and the formation of public policy 10. Explain, take and defend positions regarding the relationship of the United States to other nations and to world affairs 11. Explain, take and defend positions, and evaluate the role of citizenship and the rights and responsibilities of citizens 12. Explain, take and defend positions and evaluate issues related to the difference between political and social participation and responsible participation in American democracy

(continued)

Table 3.2. (continued).

Content Area	Standard
Geography	1. Know and understand how to use maps and other geographic representation, tools, and technologies to acquire, process, and report information 2. Use mental maps to answer complex geographic questions, make generalizations based on patterns, and apply models of spatial organization to decision making 3. Know and understand the relationship of physical and human regional characteristics to the development of culture 4. Understand the four basic components of the Earth's physical system, atmosphere, biosphere, lithosphere, and hydrosphere and how they relate to the formation of an ecosystem 5. Understand the nature and interdependence of the Earth's human systems as they relate to migration patterns, economic interdependence, control, and human settlement 6. Understand the relationship between the environment and society 7. Understand how to apply geography to interpret the past, explain the present, and predict the future
Foreign language	1. Use a target language (other than English) to establish and maintain personal relationships with a native speaker 2. Use target language to obtain, process, and provide information and for leisure and personal enrichment 3. Gain knowledge and understanding of culture's belief system, art forms, and society and explore academic, personal, and historic areas of interest 4. Access information through target language and use knowledge in integrated ways 5. Use knowledge of target language to better understand own language and culture 6. Use target language as a mechanism to participate in multilingual communities and a global society
Arts	1. Understand and explain dance elements and movements as a mechanism to communicate, explain cultures and historical periods, make connections to healthful living and other disciplines, and as a demonstration of critical and creative thinking 2. Sing, perform, improvise, compose, arrange, read, listen to and evaluate music 3. Understand the relationship of music to other art forms and to other disciplines 4. Understand the connection between music, history, and culture 5. Write scripts, act, design and produce, research and evaluate plays as they relate to personal experience, heritage, culture, and history

Table 3.2. (continued).

Content Area	Standard
Arts (continued)	6. Analyze, compare, and integrate various art forms 7. Understand the relationship of art forms to the past and present 8. Use visual arts to understand, reflect, and make connections among disciplines
Health	1. Analyze, describe, and explain concepts related to health promotion and disease prevention 2. Access, analyze, and evaluate health information and health promoting products and services 3. Develop health-enhancing behaviors 4. Analyze and evaluate behaviors to reduce health risks 5. Analyze and evaluate the influence of culture, media, technology and other factors on health 6. Demonstrate effective communication skills to enhance health 7. Use goal-setting and decision-making skills to enhance health 8. Demonstrate the ability to advocate for better personal, family and community health
Physical education	1. Participate in a variety of physical activities 2. Demonstrate intermediate or advanced skills in at least one physical activity 3. Apply scientific knowledge and principles to the improvement of motor skills related to physical activity 4. Design, monitor, and maintain a physical fitness program 5. Identify the benefits and costs of regular physical activity 6. Demonstrate responsible personal and social behavior when engaging in physical activity 7. Understand that physical activity enables one to communicate with a variety of people and leads to challenge, self-expression and enjoyment

Source: Adapted from "Struggling for Standards," 1995.

general understanding of the scope of the standards being established.

A PARADIGM SHIFT

Focusing on standards such as the ones summarized in Table 3.2 may result in propelling the United States into a new parading for international education since these standards are quite similar to those already in place in many other countries. In 1993, the Center for Education Management in the Netherlands conducted a study to determine the characteristics of education for the year 2010 (Kobus and Toenders, 1993). Using a limited Delphi technique, educators built scenarios to describe education for children and adults in the following areas: content, learning environment, learning and professional life, the teacher, the teacher student relationship and technology. The themes and patterns emerging from these scenarios are quite similar to the learning environment necessary to support achieving the emerging standards in the United States. These are summarized in Table 3.3 (pp. 10–17).

Table 3.3. Projected characteristics of education in the year 2010.

Area	Emerging Themes and Patterns
Content	While content will remain an integral part of all instruction, there will be a shift toward the following characteristics: • more process-oriented, less content-oriented • more emphasis on skills, less emphasis on knowledge • information skills: collecting, evaluating, processing • problem-solving skills • research skills • communitive skills: teamwork, negotiating • project-oriented (communal) activities • personality development: choosing, focusing one's attention, committing oneself • more international • more technological • more determined by the labor market
Learning environment	The learning environment will expand beyond the school site. Although schools as sites of learning will not disappear, they will change in the following ways: • decreasing importance of classroom as site of learning • more distance learning, tele-learning • more on the job learning • more learning takes place at home • more learning takes place at local resource center, such as libraries, community centers, museums and telecommunication centers • colleges will be learning laboratories with very diverse means and opportunities for learning (learning resource centers) • colleges will be electronic nodes which open up worlds of knowledge • having good libraries is increasingly significant • accessibility increases: open more hours a day and more days a year • more emphasis on learning skills in practical situations

Table 3.3. (continued).

Area	Emerging Themes and Patterns
Learning environment (continued)	<ul><li>more emphasis on doing assignments in groups</li><li>more emphasis on social interaction as part of the learning process</li><li>groups of students will be more heterogeneous in terms of age, ethnicity, and gender</li></ul>
Learning and professional life	There will be less of a distinction between the school and the workplace as the school environment blends with the work environment. Adults and children will come to accept the concept of life-long learning as an expected component of adulthood. The role of teacher will be broadened to include the traditionally certified adult and the adult worker with expertise to share. There will be an increase in the collaborative relationships among the school, business and community. Not only will the teacher's role change from dispenser of knowledge (sage on the stage), teachers will have to acquire new attitudes and skills.
Teacher	Role characteristics:<ul><li>facilitator of learning</li><li>coach and stimulator</li><li>collaborative participant</li><li>assessor</li><li>reader of computer-generated test results</li><li>adviser on learning trajectories</li><li>progress administrator</li><li>information organizer</li><li>guide in the use of information</li><li>provider of basic assistance for using technology</li><li>"linkage pin" between marketplace and college</li><li>creator of new knowledge</li><li>curriculum developer</li><li>courseware developer</li></ul>Attitudes and skills:<ul><li>a learner's attitude, teachers need to go on learning constantly</li><li>capability to work in a team</li><li>counseling skills (individuals and groups)</li><li>being imaginative, adaptable, stimulating, trustworthy</li><li>evaluative skills, testing skills</li><li>well oriented in professional field</li></ul>
Teacher/student relationship	Students and teachers will increasingly engage in a supportive, collaborative, collegial relationship where learning contracts are negotiated based on individual student needs, strengths, and learning styles.<ul><li>teacher and student will enter a contractual relationship</li><li>students will also have contracts with colleges</li><li>teacher and colleges provide students with services</li><li>the student is a client who regularly attends</li></ul>

(continued)

Table 3.3. (continued).

Area	Emerging Themes and Patterns
Teacher/student relationship (continued)	• the role of teacher and student may shift to enable students to be the teacher at times • the lecture disappears and is replaced by interaction • more virtual interaction and communication through networks the learning environment • more use of audio, video, courseware, CD-I • more often individual contact between teacher and student • the use of networks and multimedia will dramatically change
Technology	Major technological innovations will include: • wireless communication • interactive multimedia • voice recognition • handwriting recognition One small technological device will perform many functions: • computational power • electronic textbook • electronic notebook • multimedia interface • electronic library, database search tool • electronic mail terminal • telephone/fax/low-resolution videophone

NEW STANDARDS PROJECT: LINKING STANDARDS TO ASSESSMENT

The New Standards Project is designed to develop a system of assessments based on world-class standards of student performance. These standards are directly linked to the standards developed by the national content-area groups. They are meant to guide decisions on what to assess and to establish benchmarks for determining how good is good enough (*New Standards,* 1995). Work completed and underway by the National Council of Teachers of English, the International Reading Association, the American Association for the Advancement of Science, the National Science Teachers Association and the National Council of Teachers of Mathematics has been utilized by the New Standard's Project in setting standards for mathematics, language arts, and science.

Standards have also been set for a category called applied learning. "Applied learning focuses on the requirements for effective participation in the emerging forms of work and work organization characterized by high performance work places" (*New Standards,* 1995, p. 2). Work of the Secretary's Commission on Achieving Necessary Skills (SCANS) and other related international efforts has provided the foundation for producing the applied learning standards.

The applied learning standards are divided into two groups, problem solving and tools and techniques. There are five categories of problem solving standards and three categories of standards classified as tools and techniques (*New Standards,* 1995):

- problem solving—project design
- problem solving—planning and organization
- problem solving—teaching and learning
- problem solving—meeting client needs
- problem solving—improving the system
- communication tools and techniques
- information technology, tools and techniques
- teamwork

Many of the applied learning standards have been integrated into the standards for math, science, and language arts. For instance, a project designs standard incorporated into math and science could be "design and manufacture building blocks for use in elementary classrooms" or design a weather station for the school and produce daily school weather reports (*New Standards,* 1995, p. 346).

The critical difference between the standards developed by the associations and the ones developed by the New Standards Project is their purpose. The content standards are designed to promote the development of rich programs and provide a tool for checking their quality (p. 2). These standards represent a full range of the content that might be included in a program. The New Standards Project attempts to define those performance standards that are essential and can be assessed utilizing internationally established methods. The New Standards assessment system includes two components: "portfolios of work demonstrating performances produced by students over extended periods of time and with opportunities for revision;

and examinations (known as reference examinations) completed under on-demand conditions" (p. 4).

By establishing content standards and performance assessment standards that engage teachers and administrators in the examination of student work over time, students are engaged in constant learning about the nature of high-quality work, about themselves as learners and workers, and about the phenomena they are studying (Darling-Hammond et al., 1995, p. 252).

REFERENCES

Curry, Brian & Temple, Tierney. (1992). *Using curriculum frameworks for systemic reform.* Alexandra, Virginia: Association for Supervision and Curriculum Development.

Darling-Hammond, Linda, Ancess, Jacqueline, and Falk, Beverly. (1995). *Authentic assessment in action.* New York: Teachers College Press.

Diegmueller, Karen. (June 21, 1995). Two independent panels to review controversial U.S., World History Standards. *Education Week,* p. 9.

Gandal, Matthew. (1994). *What college-bound students abroad are expected to know about biology.* Washington, D.C.: American Federation of Teachers.

Gorman, Joseph T. (June 21, 1995). Harsh reality at graduation season: Schools are getting better—but too slowly. *Education Week,* pp. 51, 60.

Kobus, Marc & Toenders, Liny. (1993). *School 2010.* Arnhem, the Netherlands: Interstudie, Center for Education Management.

Lancaster, Laura & Lawrence, Leslie. (1992–93). *Handbook for local goals reports: Building a community of learners.* Washington D.C.: The National Education Goals Panel.

Lewis, Anne. (June, 1995). An overview of the standards movement. *Phi Delta Kappan,* pp. 744–750.

The National Education Goals Report. (1994). *Building a nation of leaders.* Washington D.C.: National Education Goals Panel.

New standards. Draft 5.1. Washington D. C.

Porter, Andrew. (January–February, 1995). The uses and misuses of opportunity-to-learn standards. *Education Researcher.*

Smith, Marshall S., Furhman, Susan H., & O'Day, Jennifer. (1994). National curriculum standards: Are they desirable and feasible? *The governance of curriculum: The 1994 ASCD yearbook.* Elmore, Richard & Fuhrman, Susan (Eds.). Alexandria, Virginia: Association for Supervision and Curriculum Development.

Statewide education reform survey. (1994). Frankfort, Kentucky: The Kentucky Institute for Education Research.

Steffy, Betty E. (1993). *Kentucky education reform: Lessons for America.* Lancaster, PA: Technomic Publishing Co., Inc.

Steffy, Betty E. (1995). *Authentic assessment and curriculum alignment: Meeting the challenge of national standards.* Rockport, Massachusetts: Pro>Active Publications.

Struggling for standards. (April 12, 1995). *Education Week.*

CHAPTER 4

Toward Balanced Assessment

ASSESSMENT practices to document student achievement in the United States are evolving. While teacher-made tests tend to remain traditional, there is significant interest at the state level in assessment practices that include more authentic measures (Barton and Coley, 1994). Almost all of the states are exploring ways to document student achievement of performance standards through state assessment measures ("Struggling for Standards," 1995). Gradually, these state assessment measures are incorporating more open-ended response items, writing prompts, and performance events to compliment the more traditional multiple choice items, completion and true and false statements (Steffy, 1995).

RETHINKING ASSESSMENT

Our traditional view of knowledge is one based on the acquisition of information in a particular subject area. As one becomes more knowledgeable, one acquires more information. "Knowledge in this sense is a collection of discrete instances of truths or information, identified categorically in subjects or fields that are assumed to be historically linear, progressive, and additive accumulations of concepts, truths, and information". (Delandshere and Petrosky, 1994, p. 11). The assessment of knowledge defined in this way traditionally takes the form of asking students to identify "bits" of knowledge or recognize the accuracy of knowledge when given a list of accurate and non-accurate "bits." This type of assessment is based on the judgment that there is one right answer.

The current trend toward more authentic measures of what students know and are able to do requires assessment techniques that accommodate multiple approaches to solving problems and a wide range of possible solutions. This trend is directly linked to the post-structuralist view of knowledge. Knowledge "is produced by individuals from what exists within the possibilities of their language or discourse. When thought of in this way, knowledge is not a collection of discrete instances of truths; knowledge is, rather, what people create, what they express, in discourses" (p. 12). Knowledge becomes the ability to use discrete facts to solve problems within the limitations of a language.

The traditional language of assessment protocols includes the concepts of reliability and validity. This language is used in the field of educational measurement to set the boundaries of what is acceptable practice. Assessment measures must be reliable and valid. Reliability refers to the notion of consistency of the assessment measure. Validity refers to the "extent to which test scores or responses measure the attribute(s) that they were designed to measure and have the impact that is sought" (Wheeler and Haertel, 1993, p. 150). Because of the nature of assessment tasks and scoring techniques, current changes in assessment practices have led to questions regarding validity and reliability. This is a particularly knotty problem in states where assessment results are linked to high stakes sanctions and rewards. We may be faced with a situation where the language of educational measurement and the boundaries set by that language have not yet evolved to allow for changes in how we define and assess knowledge.

STATE ASSESSMENT PRACTICES

Changes in classroom assessment practices are often related to changes in state mandated tests. In a survey of state testing practices, over forty states indicated that the purpose of the state testing program was accountability. A large number of states, thirty to thirty-nine, also use the state testing program as a measure of program evaluation and improvement of instruction. It has been estimated that at least 14.5 million students participate in state assessments each year. If the cost of scoring each of these assessments is $3.00 (some state authentic assessment systems cost as much as $20 per student), the taxpayers are spending over $43 million each year for this activity. Since many studies have documented that almost 70 percent of student achievement levels can be predicted based on the socioeconomic level of the family and the educational level of the mother, one wonders if the money is being spent well. There is no indication that states are constructing less costly testing systems. In fact, the trend is toward the inclusion of more costly, authentic measures of student achievement. The subject areas most often assessed by state programs are language arts and mathematics. Over thirty-five states include some mechanism for assessing student writing as part of the system. Many states are expanding the use of authentic assessment in areas such as science, social studies and reading as well as mathematics and writing.

While some state testing occurs at every grade level, the most frequently tested grade is eight (Barton and Coley, 1994). This is in keeping with other industrialized countries where students are tested between the ages of twelve and fifteen to determine their eligibility for advanced programs. The difference between the assessments in the United States and many other countries is quite significant. In France, for example, if a student fails to meet the entry requirements for the academic track, he/she is assigned to the professional/vocational track. The determination is based on course grades and exams. In the United States, if a student does not score well on the state assessment, it does not mean that he/she cannot gain admission to college preparatory classes in high school. The accountability movement in the United States is more related to school accountability than student accountability.

Students in fourth grade are the next most frequently assessed, and there is considerable state testing at the third, fifth, sixth, tenth, and eleventh grade. The least assessed grades are kindergarten, first, second, and twelfth. In any given year approximately 36 percent of all students are assessed. It is interesting that students graduating from our high schools do not have to pass content area state tests; this is not the case in other industrialized countries. For example, in France, in order to receive a baccalaureate and qualify for college entrance, a student would have to take and pass seven to eight exams. There is no such requirement in the United States.

ACHIEVEMENT AS MEASURED LEARNING

Achievement has been defined as measured learning. Not all learning is measured. Figure 4.1 depicts this concept.

It would be impossible to measure all the learning that takes place in schools. The idea of assessment is to decide what is the most important learning to assess and then construct measures to determine how much learning has taken place. At best, the practice of assessment provides only a partial picture of what has been learned. It is a flawed pseudo-science. The circle in Figure 4.1 represents all of the learning taking place in school. The square represents the learning that is assessed. As depicted in Figure 4.1, some of what is assessed is not learned in school. The ideal relationship between learning and achievement might resemble Figure 4.2.

In Figure 4.2, the learning that is assessed is a subpart of the

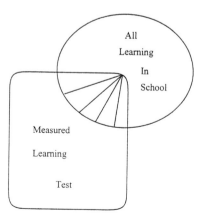

Figure 4.1. Achievement as measured learning.

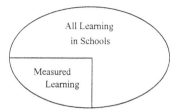

Figure 4.2. Ideal relationship of achievement to learning.

total learning. Traditional means of assessment have been typically used to measure the acquisition of basic skills such as the ability to solve mathematical problems using the four operations of addition, subtraction, multiplication, and division. Authentic assessment measures are typically used to assess the use of basic skills to solve problems. Traditional assessment measures usually call for one right answer. Authentic assessment measures enable students to come forward with multiple "right answers."

There are strengths and weaknesses attached to each type of assessment. There is no one perfect type. The most important consideration is determining what type of learning is to be assessed. Table 4.1 outlines the strengths and weaknesses of the most typical traditional and authentic assessments currently in use in the United States.

EXAMPLE: PORTFOLIO ASSESSMENT

One of the most creative uses of authentic assessment has been employed by the staff at Central Park East Secondary School in East Harlem, New York City (Meier and Schwarz, 1995). This alternative, secondary school uses a juried, portfolio review process to determine whether a student has met graduation requirements. "The final high school diploma is not based on time spent in class or Carnegie units, but on each student's clear demonstration of achievement" (p. 27).

The student population served by this school is largely African American or Latino and most come from the East Harlem area. Yet the staff estimates that 97.3 percent of the students who attend Central Park East Secondary School (CPESS)

Table 4.1. Strengths and Weaknesses of Traditional and Authentic Assessments.

Type of Assessment	Strengths	Weaknesses
Criterion referenced	1. Assess specific skills 2. Assesses whether a student has acquired a specific knowledge base 3. Usually accompanied by a skill list by which the test was constructed 4. There is one right answer to each question 5. Lends itself to item analysis 6. Establishes reliability and validity more easily than with authentic measures 7. Usually uses a multiple choice format 8. Easy to score	1. Does not allow for a variety of responses 2. Permits guessing 3. Generally assesses the lowest levels of learning such as recall of basic facts 4. Does not permit a student to demonstrate all that he/she knows about a topic 5. Generally does not require students to solve problems through critical thought
Norm referenced	1. Enables comparisons among individuals or groups 2. Easy to score 3. Cost effective	1. Results predicted by SES 2. Test items not tied to any particular curriculum 3. Teaching to the test is considered unethical 4. Assumes that a teacher is not a purposive variable 5. Demands results where there are winners and losers 6. Demands failure
Writing sample	1. Assesses a student's ability to write a short, concise, well-constructed response to a writing prompt in a timed situation 2. Requires the ability to organize ideas 3. Provides for multiple responses	1. Takes a long time to score 2. Sometimes questions rater's reliability 3. Scoring procedures allows for more subjectivity 4. Costly

Table 4.1. (continued).

Type of Assessment	Strengths	Weaknesses
	4. Assesses a valued, fundamental skill 5. Requires the application of a variety of basic skills such as the ability to read, use proper grammar, understand correct punctuation and capitalization and display the use of figurative language	
Open-ended response item	1. Requires critical thinking 2. Requires problem solving 3. Permits multi-step responses 4. Used more often in other industrialized countries 5. Enables a student to use his/her knowledge base to solve the problem 6. Enables responses to be judged at various levels of performance 7. Offers no one right answer	1. Difficult to assess 2. Generally requires a scoring rubric for each item 3. Expensive to grade. 4. Criticized by religious right when item deals with taking a position involving values 5. Offers no one right answer 6. Validity and reliability problems
Performance event	1. Can assess the ability of students to work together to complete a task 2. Can assess the ability of a student to create a product or perform a concrete skill such as use word-processing equipment to publish a story 3. Offers no one right answer 4. Can assess integrated learning 5. Can assess multi-step problem-solving	1. Difficult to assess individual work if it is a group activity 2. Reliability and validity difficult to achieve 3. Group-work sometimes opposed by the religious right 5. Costly

(continued)

Table 4.1. (continued).

Type of Assessment	Strengths	Weaknesses
Portfolio	1. Provides for a collection of a student's best work 2. Engages the student n assuming ownership for what goes into the portfolio 3. Provides a record of a student's growth over time. 4. Enables students to capitalize on learning style strengths in choosing the format and medium for materials included in the portfolio 5. Permits revision and editing of work included in the portfolio 6. Encourages student decision making regarding content to be covered	1. Difficult to assess 2. Requires the development of a scoring rubric that assesses level of performance within a structure of specific standards 3. Takes time to develop thus reducing the amount of instructional time available for specific content-focused teaching 4. Requires additional storage capability

graduates from high school and 90 percent of the graduates attend college.

The portfolio review process utilized by CPESS is not an add-on. Rather, it is integrated into the instructional program so that students are involved in the preparation of their portfolios over a long period of time. Fourteen portfolios are required and each must demonstrate a student's ability to provide evidence for what he or she knows.

Students must demonstrate that they understand the perspective or point of view exhibited by the knowledge, are able to make connections between this knowledge and the larger knowledge base, can speculate about how different events might impact the way things are, and understand the relevance of the knowledge. Five key questions guide students' inquiry (Meier and Schwarz, 1995, p. 30):

1. How do you know what you know? (Evidence)
2. From what viewpoint is this being presented? (Perspective)
3. How is this event or work connected to others? (Connections)
4. What if things were different? (Supposition)
5. Why is it important? (Relevance)

The scoring rubrics used by the Graduation Committee are called "scoring grids" and each discipline has developed its own. These scoring grids are periodically reviewed and updated to ensure that advances in the skill and knowledge base of a particular discipline are incorporated into portfolio requirements. These portfolios are expected to represent each student's best work. While each student selects one of the portfolios and presents it as a final senior project, all graduates are required to make major presentations in the four areas of science/technology, mathematics, history and social studies, and literature. In addition, each student may select three additional presentation areas from the following list (pp. 31–32). These seven areas are considered the student's majors.

- ethics and social issues
- fine arts/aesthetics
- practical skills

- media
- geography
- language other than English
- physical challenge
- school and community service
- autobiography
- postsecondary plan

Letter grades are not used in the assessment. Rather, the Graduation Committee assigns the student one of four levels of achievement for the seven major areas: Distinguished, Satisfactory Plus, Satisfactory, or Minimum Satisfactory. The other seven areas, which are called minors, receive a pass/fail designation.

The staff at CPESS believes that this portfolio review process enables graduates to more clearly demonstrate what they know and are able to do. In the process of developing the portfolios students make critical decisions about what knowledge is most important for them to know. This is a role traditionally held by the teacher. The CPESS model requires students to take responsibility for their own learning and enables teachers to assume the role of facilitator. At CPESS the student is a major player in creating the learning/achievement relationship.

EXAMPLE: OPEN-ENDED RESPONSE ITEM

Open-ended response items required for both the organized recall of knowledge and the interpretation of information are routinely included in the assessments incorporated into the baccalaureate exams in France. The example cited here is adapted from the 1992 biology exam (as reproduced in Gandal, 1994). This test is developed primarily by the Ministry of National Education. "The Ministry determines the topics to be covered in each year's examinations, as well as the dates and administrative procedures for the tests" (p. 34). Students are permitted to choose between two sections and are allotted three hours to complete the exam. One section is shown here to provide the reader with an example of a multi-level open-ended response item calling for not only a demonstration of a sophisticated knowledge base, but also calling for the application of high level critical thinking skills. While the graphs included in the question are not shown here, the reader can clearly discern the skills required to successfully complete this exam.

Section I

Part A: Organized Recall of Knowledge (10 points total)

Measurements taken from a cell culture during the interphase preceding mitosis have revealed the following:
- a doubling of the quantity of DNA in the nucleus
- an increase in the weight of cytoplasmic proteins

Using carefully selected, concise, and clearly annotated diagrams, illustrate the two processes involved in preparing for mitosis. Limit your answer to the stages that take place in the nucleus. Without detailing the mechanism of protein synthesis, explain how these two processes prepare the cells for preserving:
- their genetic information
- their protein mass from one generation to the next

Part B: Interpretation of Documents (10 points total)
Effect of two hormones on their target cells

We want to study the action of a hormone, insulin, on fat cells and the action of a gastric hormone, bombazine, on pancreatic acinar cells, which secrete pancreatic juices.

Document 1

Experiment A

Fat cells from a rat are placed in a culture medium containing glucose marked with radioactive C. The radioactivity of the CO_2 produced by the fat cells as a function of the concentration of insulin added to the culture medium is measured shown in graph not reproduced here.

Experiment B

Insulin marked with a radioactive amino acid is injected intravenously in a mouse; we observe that the plasma membranes of the fat cells are radioactive.

Document 2

Experiment C

Pancreatic acinar cells are placed in a culture medium. The rate of secretion of digestive enzymes released by the cells in the medium is measured as a function of the concentration of bombazine added to the culture medium. (Graph illustration not included here.)

Experiment D

In the presence of bombazine in a culture medium, an increase in the concentration of Ca^{2+} is observed in the cytoplasm of the cells in the culture.

Experiment E

In the absence of bombazine in this culture medium, a release of digestive enzymes by exocytosis is observed when Ca^{2+} is injected into the cytoplasm of an acinar cell.

Question

1. Based on arguments drawn from Documents 1 and 2 show (2 points each):
   - the effects of insulin on fat cells and the effects of bombazine on acinar cells
   - that these cells are target cells responding to a hormone message that you will define (Your answer should include an explanation of what a hormone message is.)

Triggering the Secretion of Insulin

Insulin is synthesized by the pancreas. Under normal physiological conditions, insulin secretion increases when the concentration of glucose in the blood rises.

Document 3

Experiment F

An islet of Langerhans isolated by microdissection is preserved under conditions ensuring that it will retain its physiological integrity. The penetration of calcium into the B cells of an islet of Langerhans is measured at 5-minute intervals shown in graphs not included here, and the secretion of insulin by the same cells is measured every minute as a function of the concentration of extracellular glucose.

Experiment G

The injection of Ca^{2+} in the cytoplasm of the B cells of the islets of Langerhans stimulates exocytosis of insulin, even in the absence of glucose.

Question

2. What information can you derive from a side-by-side comparison of all the data provided in Document 3 on triggering the secretion of insulin? (1 point)

Document 4

It has been shown that the membrane of the B cells contains calcium ducts that are dependent on the transmembrane voltage; when these open, calcium penetrates the cells.
The difference in the transmembrane potential of B cells is measured as a function of the concentration of extracellular glucose.

Questions

3. What new information does Document 4 provide? (2 points)
4. Use all your answers from Part B to draw a functional diagram showing the chronology of events leading to exocytosis of insulin.

This example was chosen because it provides the reader with an understanding of how open-ended response items can require not only a demonstration of content knowledge, but how they enable a student to use that knowledge to solve problems, take positions, or explain events. Open-ended response items, such as the one shown here, require students to use an array of critical thinking skills—from recall to analysis, synthesis, and evaluation.

EXAMPLE: PERFORMANCE EVENT

Like portfolios and open-ended response items, performance events may take many forms. The example provided here was developed by an unknown Kentucky teacher shortly after the Kentucky Education Reform Act (KERA) was passed.

Description of Assessment Task

Making a decision about a college or career is a major decision in a person's life. The following task is designed to allow the student to demonstrate that he or she is capable of researching this important topic and developing relevant conclusions. Students will select a career, research the career, conduct one or more interviews, and make a presentation to the class. They will also develop a personal file for pursuing that career.

Part I

Your team is to brainstorm possible careers from which each team member will select a career to explore and justify the choice. Each team member is to write down all that he or she already knows about the career and to identify for the group information categories requiring more investigation. These may include: 1) nature of work, 2) working conditions, 3) employment outlook, 4) job availability, 5) salary, 6) education and training requirements, 7) opportunities for advancement.

Part II

Each team member will develop a research paper for presentation. Documentation for the presentation must include:

1. Note cards and bibliography cards
2. Questions used in the interview process
3. Outline of information
4. All rough drafts and edited copies of the report
5. A final paper, complete with endnotes and bibliography

Note: Critiques and editing must be done by at least two different team members.

Resources may include the *Occupational Outlook Handbook, Kentucky Occupational Guide,* and career encyclopedias. Interview opportunities include actual career persons, a visit to Job Services or any university, technical or vocational school.

Following the research and interview process, each team member is to write and present a four- to six-page paper that includes reasons for the career choice, pertinent information about the career, and personal conclusions about the career for that team member. A presentation of the research findings is to be made to the class.

Part III

Each team member is to develop a personal file that can actually be used to pursue that career. This personal file must include: 1) a letter requesting a job interview, 2) a resume and 3) a completed job application or application for entry into a college or vocational/technical school.

Each component of the file must be critiqued and edited by at least two team members.

Performance Criteria

- production of ideas from team brainstorming efforts
- demonstration of research skills with documentation of steps and components for the research paper
- critiquing and editing of one or more team member's paper(s)
- critiquing and editing of student's paper by at least two team members
- production of personal file with all components

- critiquing and editing of one or more team member's personal file
- critiquing and editing of student's personal file by at least two team members

Suggestions for Scoring Rubric

Exemplary Performance

Part I

All team members contribute ideas for career opportunities. Number of ideas exceeds the number of team members. Each team member selects a career and lists ten or more items of information needed.

Part II

A research paper of six or more pages is written and presented to the class. Documentation on all components of the research is present and signed off by all team members.

Part III

A personal file is developed with all components present and signed off by all team members.

Acceptable Performance

Part I

All team members contribute ideas for career opportunities. Number of ideas equals the number of team members. Each team member selects a career and lists six or more items of information needed.

Part II

A research paper of four to six pages is written and presented to the class. Documentation on all components of the research is present and signed off by at least two team members.

Part III

A personal file is developed with all components present and signed off by at least two team members.

Unacceptable Performance

Part I

Not all team members contribute ideas for career opportunities. Number of ideas is less than the number of team members. Each team member selects a career but identifies less than six items of information needed.

Part II

A research paper of less than four pages is written and presented to the class. Documentation is missing on some components of the research and is not signed off by at least two team members.

Part III

A personal file is developed but does not contain all components and is not signed off by at least two team members.

Each of the examples presented represents an assessment strategy that requires students to apply an integrated set of skills. Completing the assessments requires a lengthy period of time and allows for student diversity in responding to the problems. Student responses to the open-ended item would be the most similar. Student responses to the portfolios would probably be the most divergent. Authentic assessments such as these should be used in combination with other more traditional assessment strategies to enable the student and the teacher to gain insight into the student's knowledge and problem-solving skills.

AUTHENTIC ASSESSMENT AS A CATALYST FOR SCHOOL REFORM

In 1989, the Education Testing Service issued a report summarizing the status of high school students' performance. The report indicated that 61 percent of seventeen-year-old students could not read or comprehend high school material. Almost half could not use mathematical operations to solve multi-step problems and less than half could evaluate the procedures in a scientific inquiry. In addition high school juniors had little sense of historical chronology, did not read much literature and were unfamiliar with the use and potential applications of computers [Educational Testing Service (ETS), 1989, p. 26].

A few high schools are using authentic assessment measures to dramatically change this dismal scenario. Two examples are provided here. Both are the result of case studies supported by a grant to the National Center for Restructuring Education, Schools, and Teaching (NCREST) at Teachers College, Columbia University, and chronicled in the 1995 publication, *Authentic Assessment in Action: Studies of Schools and Students at Work* (Darling-Hammond, Ancess, and Falk, 1995). The case studies examined how the schools' assessment strategies work, how they were developed and introduced, what problems were encountered, the effects on staff and students, what changes occurred in the classroom, and what effects the assessments had on student learning (p. 15). The assessment strategies employed were designed to measure actual performance. "They are intended to provide a broad range of continuous, qualitative data that can be used by teachers to inform and shape instruction. They aim to evaluate students' abilities and performance more fully and accurately, and to provide teachers with information that helps them develop strategies that will be helpful to the real needs of individual children" (p. 10). Each of these schools had made a commitment to the nine principles of the Coalition of Essential Schools. These principles include the following:

- Schools should help students use their minds.
- The school's agenda should be simple.

- School goals should apply to all students.
- Teaching and learning should be personalized.
- Schools' view of students should be student as worker.
- Students entering secondary schools should have basic competence in mathematics and language.
- School community should be accepting and trustful.
- Teachers and principal should perceive themselves as generalist.
- Student load should not exceed eighty students.

INTERNATIONAL HIGH SCHOOL

This high school serves students who have been in the United States less than four years and score below the 20th percentile on English language proficiency tests (Darling-Hammond, Ancess, and Falk, p. 115). In 1988, an article appearing in the *New York Times* indicated that among the 310 students attending, thirty-seven nations were represented and thirty-four languages were spoken. All seniors graduating that year (fifty-four) planned to attend college. How did the school achieve these results? "The answer to this question lies in International High School's commitment to a collaborative, experiential approach to teaching and learning, married to school-wide process of reflection and authentic assessment deeply embedded in all the activities of the school" (p. 115). By 1992 the high school served 459 students from fifty-four countries representing thirty-nine languages, over three-fourths of the students qualified for free or reduced lunch, and the ethnic composition of the school included 45 percent Latino (p. 116). International High School Regents competency test results indicated that 100 percent of the students passed Regents tests in mathematics, global studies, science, and occupational education, 99 percent passed reading, 98 percent passed writing, and 97 percent passed American studies. These results represent an impressive success story for any high school in New York but an especially impressive outcome for students entering with such severe documented deficiencies.

The mission of International High School is described as fostering the linguistic, cognitive, and cultural skills necessary for [students'] success in high school, college, and beyond (p. 117). Some of the unique characteristics of this high school include

- seventy-minute periods
- conceptual theme focused classes (e.g., interdependence, causality)
- interdisciplinary instruction
- extended-day tutorials
- located on a community college campus
- half-day student internships during one trimester per year
- four classes during each thirteen-week cycle
- collective staff planning (one-half day per week)
- teacher portfolios

The student evaluation system has evolved over time from a more traditional one composed of periodic quizzes to one built on the concepts of "a continuous process of self-reflection, peer assessment, and teacher assessment, organized around collaborative performance tasks and individual portfolio development" (p. 119). Participants in the program have identified three factors that contributed to the development of the current authentic assessment system. The first deals with the creation of a learner-centered instructional model; secondly, the increasing dissatisfaction of teachers with the more traditional assessment practices; and finally, the evolution of a new staff evaluation system developed by the faculty.

"Collaboration between and among students and faculty is at the core of the learning environment (p. 120). Students are placed in heterogeneous, interage groups. Each group of approximately seventy-five students works primarily with four teachers within a learning community characterized by collaboration, active learning, whole language and authentic assessment. "Teachers plan together and students work extensively in small groups with teachers alternately coaching, assessing, questioning, and prodding them; with their groups, students work on individual as well as group tasks" (p. 122).

Over time, the teachers grew increasingly dissatisfied with traditional assessment practices. They proved inadequate for measuring the growth students were experiencing in both written and spoken language. Project work was added as a mechanism for assessing what students know and were able to do. The projects involved both individual and group work. The faculty came to believe that traditional assessments did not "provide students with the independence and collaborative work skills that translate into successful, college-level work" (p. 123). Over time there was acceptance of the idea that when students are given responsibility for their own learning, they accept that responsibility for the outcomes of that learning.

A key factor in the development of the student evaluation system involved the development of a new teacher evaluation plan. As the process evolved, teachers "began to appreciate the power and potential of collaborative problem solving, self- and peer-assessment, and exhibitions of their work for enriching their professional learning and development" (p. 126). In the end, the faculty review process took on many of the characteristics of the student review process. These included encouraging experimentation, providing time for sharing ideas, insights, and self-reflection, exposing staff to cutting-edge pedagogy, and "enabling them to develop a philosophy of what constitutes effective teaching and counseling" (p. 126). In this way the staff began to embrace "an understanding of and commitment to ongoing assessment, self-directed learning and assessment, evaluation of both process and product, and evaluation from multiple perspectives by multiple colleagues" (p. 126).

Assessment Strategies

Self-assessment and self-improvement form the bedrock of the evaluation system used at International. Three of the programs, Beginnings, the Personal and Career Development/Internship Program, and Motion rely heavily on authentic assessment measures to validate student achievement. All three programs utilize the following components (Darling-Hammond, Ancess, and Falk, 1995, p. 128):

- performance-based assessment produced while work is in progress

- summative evaluation based on multiple dimensions
- assessment of group and individual work
- assessment of process and products
- evaluation from multiple points of view

Ongoing assessment in the Beginnings program is based on group work, exhibitions and coaching. The final evaluation for this program is done with a conference. The process being evaluated is group work and the product being evaluated is an autobiography. Staff reported that no student ever lost his/her autobiography. The evaluation of this program is conducted by the student, peers, and the teacher.

This program is a consolidation of three courses: biology, social studies and English as a Second Language. It is completed by newly-enrolled students. The autobiography developed by students includes "chapters on their childhood, on life in their countries of origin, on their milestones, their immigration, their responses to the United States, and self-assessments of their interests, abilities, and skills in the context of career exploration" (p. 129). Students work in collaborative groups that enable them to

- learn alternative work habits
- develop peer assessment skills
- acquire an "internal standard" for the quality of their work
- develop a self-motivated work ethic

Formal assessment takes place two times during the Beginnings program. The first assessment is completed in small groups where students make presentations of their autobiographies. Each student in the group gives and receives feedback about his/her autobiography. For the summative evaluation, the student meets with three Beginnings teachers and two peers the student has selected. Students are expected to display critical thinking skills during these presentations. In addition to identifying their best work, they are expected to be able to explain connections and identify and analyze the elements of their work. The final conferences are held in English. As part of the final conference, students evaluate the effectiveness of the course for faculty.

The Personal and Career Development Program extends over three cycles. Students are expected to complete a work internship four days a week and attend a school seminar one day a week during each cycle. The school seminars focus on the development of socialization and communication. "The kids learn to negotiate, mediate, to defend their positions. . . . Most of them practice these skills at their internship sites. They learn—and we teach them—the importance of being assertive without being aggressive if they feel they are being used unfairly" (p. 132).

During this program the students construct internship albums composed of four chapters. Development of the chapters requires students to

- engage in analytic self-assessment
- acquire knowledge and skills
- identify personal and career objectives
- articulate rationale for internship choices
- describe job interviews, job duties, and roles within the workplace
- interview co-workers and supervisors
- describe employment opportunities, educational requirements, and general benefits of the career area being explored

During this program, learning and assessment are interactive. The final summative evaluation involves self, peers, and faculty, and students are given letter grades ranging from A to C. Expectations for what constitutes each letter have been carefully constructed and take into consideration the level of interest, clarity, degree of detail, supportive evidence, and creativity of the presentation as well as the content and completeness of the presentation. In addition, workplace supervisors rate each student on a five-point scale utilizing nine indicators: "attendance, promptness, quality of performance, dependability, cooperation with both co-workers and supervisors, ability to learn, initiative, and growth" (p. 133). Students also complete a questionnaire assessing their internship performance and the quality of their albums.

Assessment in the Motion program consists of debriefings, two portfolio reviews, and the final conference. "The use of assessment to drive collaborative learning turns out to produce one of the most powerful experiences Motion students have" (p. 134). During this program students focus on content knowledge by working in groups to design experiments and solve problems in physics and mathematics, interpret literature, write to and with each other, and conquer physical challenges. Through these activities, students are forced to use knowledge. So that students learn to work with a variety of students in a variety of ways, students are forced to sign up to work with other students they do not know or have not previously worked with. At debriefings, each student is expected to demonstrate his/her acquisition of knowledge and skills. However, students are given a group score. This encourages the students to learn to work together and take responsibility for each others' success, to persevere. The debriefings provide a "moving picture of student work."

Portfolios provide a snapshot of student progress. They are reviewed at mid-cycle and at the end of the cycle. The Motion portfolio has four parts (p. 140):

- a data summary and samples of students' work
- personal statements: a self-assessment essay
- mastery statement
- self, peer, and faculty evaluation: assignment of grades

Students are required to report attendance, tardiness, the number of activities completed, and the titles of work samples they have included in their portfolios at mid-cycle and at the end of the cycle. This cycle includes as many as sixty activities in physics, mathematics and language arts. In students' personal statements, they reflect on the progress they have made personally and in work groups in six areas:

- language and communication skills
- individual work and responsibility
- group work and participation
- work with adults
- academic growth
- overall progress

Development of the mastery statement is an opportunity for

students to demonstrate the degree of mastery they have achieved. An example of the questions one student responded to in her mastery statement follows (pp. 147–151):

1. What have you learned about how to write well from your writing experiences in literature?
2. How does the work you have done in literature connect to the work in math and physics? Give examples from at least two of the activities you have done since mid-cycle.
3. Consider Newton's laws. Play with them in your mind. State them in simple language. Give examples. What would a world that does not obey Newton's laws be like?
4. We have been using graphs to help us understand relationships. One example is the graphs of distance, velocity and acceleration for objects in free fall. Draw and explain the graphs.
5. In several of the math activities we have drawn graphs and expressed the relationship in algebraic equation. Draw a graph with a linear relationship between two variables. Write the equation for it. What does y-intercept and the slope represent?
6. Do you think your work in project adventure changed the way you worked in class? Did you find out anything surprising about yourself?

Peers, students and faculty all participate in the evaluation process for Motion. These procedures are guided by systematic guidelines, however, they do take into consideration the individual differences and needs of the students. Twelve indicators are used to determine the final grade (p. 152). Indicators used to judge classwork include

- number of absences and number of times the student was late
- amount of work completed
- working with others
- concentration
- gives specific examples
- shows what the person has learned

- is well organized
- is neat and easy to read
- explains the connection between classes

"There is no division at International between academic and vocational skills. Critical thinking, goal setting, accurate self-assessment, effective communication and relationships with peers and supervisors, collegial collaboration, self-motivation and initiative, and application of learning to new contexts—all features of the assessments—are valuable in both the school and the workplace" (p. 166). At International the "test is the work."

THE BRONX NEW SCHOOL

Opened in 1988, the Bronx New School currently has a student population of approximately 250 students in grades kindergarten through six. The school is designed to be "learner-centered." It is "organized into heterogeneous, multi-age classes and structured to encourage and enhance collaboration among faculty, students, and students' families" (Darling-Hammond, Ancess, and Falk, 1995, p. 205). Instruction is interdisciplinary and based on the constructivist idea of active learning where students have ample opportunities to pursue and display their own interests and strengths. Enrollment in the school is by lottery and the student population consists of one-third Latino, one-third African-American, and one-third other. The school is led by a teacher-director. The intent of the school is to "provide a setting that engages learners, seeks to involve each person wholly in mind, sense of self, sense of humor, range of interests, interactions with other people in learning: that suggests wonderful ideas to children... different ideas to different children... and that lets them feel good about themselves for having them" (Duckworth, 1987, pp. 1, 7, 134, as reported in Darling-Hammond, Ancess, and Falk, 1995, p. 207).

Like the assessment system for International, the Bronx New School utilizes a variety of assessment techniques over the course of the elementary program. These techniques focus on both students' accomplishments and students' special strengths. The assessment system can be characterized as:

- documenting student growth over time
- involving families
- offering opportunities for reflection
- setting high accountability standards
- inquiring into how children learn
- encouraging flexibility in method and timing
- linking assessment to standards and indicators
- establishing benchmarks that take into account the uneven growth of young children
- designing procedures to assess a wide range of understandings

Portfolios are used as the repository of documented student growth. They include examples of student work, teacher reflections, and student comments. Portfolios travel with students over time. Entries are made at the beginning, middle, and end of each school year. The portfolio becomes the basis of interaction with parents and students and contains teacher-kept records, student work samples, and student-kept records.

> As the program grew and changed over time, teachers' beliefs about assessment changed. As reported for one teacher, "she changed her thinking and her teaching as well. She came to realize that a classroom limited only to traditional forms of academic expression excludes different types of children as well as different types of knowledge. She became poignantly aware that children who have diverse strengths and interests often feel that because there is no room in school for the kinds of activities they value, there is literally no room for them either." (Darling-Hammond, Ancess, and Falk, p. 221)

Teacher-kept records included many of the items that traditionally might be placed in a students' permanent record file. Traditionally, this was a file seen only by school personnel and occasionally shared with parents in a formal conference. By including these records in the students' portfolios, they are public entries that enable the students to better understand their own learning. These records include:

- student observations
- inventories and checklists of student skill development
- student/teacher conference notes

Teachers were free to develop their own mechanism for recording the linguistic, logical, numerical, musical, artistic, bodily, spatial, and social strengths of children. "The commonality they were seeking was not in the specifics of the instruments they used, but rather in the ways they looked at children and, subsequently, supported student learning" (p. 210). As teachers became more sophisticated in their note taking, they began to take on the role of the researcher, "jotting down" and reflecting about the learner.

Student work samples were chosen collectively by the student and the teacher. These work samples could include video, photos or descriptions of projects, and audio tapes, as well as written expression in the form of journals, narratives, and drawings. Entries were dated to show the progression of work over time. At the end of each year the portfolios were reviewed to be sure they contained a broad array of entries for all content areas. Files were purged annually so that they included approximately twelve to twenty items that best documented student growth over time and the rest were sent home to the families.

Student-kept records became a critical part of the portfolios. These included records of what the students read, projects they completed, reflections regarding conferences, attendance and disciplinary actions, logs and checklist of accomplishments, and anything else the students valued as documentation of progress. The student-kept records promoted student ownership for their own growth and development. This was not the teachers' view: it was the students' demonstration of accomplishment.

Teachers found the time necessary to manage this assessment system by extending the lunch hour and recess time each Friday. The director, assisted by other adults, provided for the students and freed the teachers to have two hours block of time for assessment-related tasks. In addition, since the teacher's role changed to more of a facilitator, there were opportunities during formal class time for note taking and recording reflective comments. On occasion, teachers would combine classes, enabling some teachers to have an additional period of time to devote to assessment. As an outgrowth of the assessment system, teachers had more opportunities to share with one another. "Together they explored themes central to understanding their teaching; posed problems; discussed dilemmas that they were finding difficult to answer; questioned knowledge they found to be problematic, defined the kind of evidence they sought in order to document and explore issues, and suggested ways they could link up diverse experiences" (p. 243).

USING ASSESSMENT TO TRANSFORM SCHOOLS

The International School and the Bronx New School are examples of the power of using the process of changing assessment practices as a catalyst for changing instructional practice. "By working on standards and assessment, looking carefully at students' work and progress, and working to develop supports for student success, teachers are engaged in constant learning about the needs and talents of their students, about the effectiveness of their teaching strategies, and about the nature of teaching and learning" (Darling-Hammond, Ancess, and Falk, 1995, p. 252). If schools establish challenging standards and authentic assessment practices, students are more likely to be engaged in striving for high expectations for skills required to function successfully in the 21st century. "By working toward challenging standards on authentic tasks, students are engaged in constant learning about the nature of high-quality work, about themselves as learners and workers, and about the phenomena they are studying. By working collectively to create and evaluate assessments, by rethinking school wide practice so that they enable students to work on and succeed at complex, extended performances, and by communicating in new ways about students' work, schools are engaged in constant organizational learning about the effectiveness of their practices" (p. 252).

REFERENCES

Barton, Paul E. & Coley, Richard J. (1994). *Testing in America's schools.* Princeton, New, Jersey: Educational Testing Service.

Darling-Hammond, Linda, Ancess, Jacqueline, & Falk, Beverly. (1995). *Authentic assessment in action: Studies of schools and students at work.* New York: Teachers College Press.

Delandshere, Ginette & Petrosky, Anthony. (1994, June–July). Capturing teachers' knowledge: Performance assessment. *Educational Researcher,* 23(5): 11–18.

Educational Testing Service (ETS). (1989). *A world of differences: An international assessment of mathematics and science.* Princeton, New Jersey: Author.

Gandal, Matthew. (1994). *What college-bound students abroad are expected to know about biology.* Washington, DC: American Federation of Teachers.

Meier, Deborah & Schwarz, Paul. (1995). Central Park East Secondary School: The hard part is making it happen. In Michael W. Apple & James A. Beane (Eds.), *Democratic schools.* Alexandria, Virginia: Association for Supervision and Curriculum Development.

Steffy, Betty. (1995). *Authentic assessment and curriculum alignment: Meeting the challenge of national standards.* Rockport, Massachusetts: Pro>Active Publications.

Struggling for Standards. (1995, April 12). *Education Week.*

Wheeler, Patricia & Haertel, Geneva. (1993). *Resource handbook on performance assessment as measurement: A tool for students, practitioners, and policy makers.* Livermore, California: The Owl Press.

CHAPTER 5

Program Evaluation

EVALUATING the effectiveness of programs is central to the continuous improvement of schools. In keeping with the concept of value-added benefit and continuous improvement, program evaluation procedures are becoming increasingly popular in schools. The purpose of program evaluation is to provide a basis for informed decision making, to capitalize on successful strategies and minimize or eliminate ineffective practices. Evaluations enable administrators to analyze program structures and implementation plans within the political and social context in which the program operates (Fink, 1995). Program evaluation, done well, can lead to continuous improvement, focus valuable resources, build capacity, empower participants, promote ownership and lead to the development of system-wide pride and staff recognition. Without effective program evaluation procedures, school personnel are unable to document "what works."

During the late 1960s and early 1970s most program evaluation efforts were focused on large-scale curriculum efforts (Herman, Morris and Fitz-Gibbon, 1987). These studies were basically quantitative in nature and relied on "experimental methods, standardized data collection, large samples, and the provision of scientific, technical data" (p. 9). It was assumed that clear cause-effect relationships could be determined and these would inform policy makers about the most effective policies to enact in order to improve student achievement. Most of these studies did not take into account contextual factors that impact the success of a program. Over time, program evaluation models began to be responsive to the unique characteristics and processes within the context of the program implementation. There is a growing recognition of the complexity of well-defined program evaluations.

PROGRAM EVALUATION STANDARDS

During the past two decades there has been growing interest in establishing and using program evaluation standards. In 1974, a committee was appointed under the auspices of the American Educational Research Association, the American Psychological Association, and the National Council on Measurement in Education to revise standards for educational and psychological tests (The Joint Committee on Standards for Educational Evaluation, 1994). The work of this committee resulted in the publication of *Standards for Evaluations of Educational Programs, Projects, and Materials,* 1981.

Twelve organizations were represented on the committee that developed this publication. In 1994, The Joint Committee on Standards for Educational Evaluation, chaired by James R. Sanders, came out with the second edition. This edition involved representatives from fifteen major educational associations including the National School Boards Association, the Council of Chief State School Officers, the American Association of School Administrators and the National Education Association. The goal of the project was "to develop standards to help ensure useful, feasible, ethical, and sound evaluation of educational programs, projects, and materials" (p. xiv). Thirty standards have been identified. They are organized into four broad categories:

1. Utility standards—intended to ensure that an evaluation will serve the information needs of intended users
2. Feasibility standards—intended to ensure that an evaluation will be realistic, prudent, diplomatic and frugal
3. Propriety standards—intended to ensure that an evaluation will be conducted legally, ethically, and with due regard for the welfare of those involved in the evaluation, as well as those affected by its results
4. Accuracy standards—intended to ensure that an evaluation

will reveal and convey technically adequate information about the features that determine worth or merit of the program being evaluated

The complete list of standards is included at the end of this chapter. They are not copyrighted material, and their reproduction and dissemination are encouraged by The Joint Committee on Standards for Educational Evaluation.

GUIDELINES AND COMMON ERRORS FOR SELECTED STANDARDS

When creating program evaluation designs, the evaluator should be mindful of guidelines and common errors identified by the Joint Committee. A few of the most salient are discussed here. For a more comprehensive listing see the second edition of *The Program Evaluation Standards: How to Assess Evaluations of Educational Programs* (The Joint Committee on Standards for Educational Evaluation, 1994).

Stakeholder Identification

This standard deals with being sure that everyone affected by the evaluation is identified and there is an attempt to meet their needs. Both stakeholders and clients typically represent diverse groups with differing points of view. It is understood that the limitations of time and resources will limit the number of stakeholders and the amount of direct involvement they have in the design and implementation of the study. However, "If stakeholder identification is not done, the evaluation may become a misguided, academic exercise, the results of which are ignored, criticized, or resisted because they do not address anyone's particular questions" (p. 25). It is also important to be sure that stakeholders are not excluded because of gender, ethnicity, or language background.

Some of the common errors made in failing to meet this standard include allowing clients to restrict evaluator's contact, implying that all stakeholder information needs to be addressed, and assuming that leaders are the most important. In school systems, the board is often the body requesting, authorizing, and paying for an evaluation. Boards can restrict contact through the level of funding, the time line, and how and when the study is reported. By limiting data-gathering sources to leaders, the evaluators run the risk of a study that is too narrow in focus and biased in terms of the point of view being addressed. On the other hand, attempting to address the needs of all interested stakeholders can cause the study to become so diluted that there is no primary focus.

Evaluation Impact

Unless an evaluation is used, there is little reason for conducting it. If stakeholders have difficulty in seeing how the information in the report translates into activities to strengthen the program, the effort and expense may not be justified. An evaluation must have impact. The evaluator is in a position to make suggestions to help program participants select more cost-beneficial approaches; stop wasteful, unproductive efforts; or see the program in a different way (p. 59). The evaluator should take an active role in showing stakeholders how the findings will inform their decision making. To the extent possible, stakeholders should be involved in determining evaluation questions. Stakeholder involvement can be enhanced by providing clients with periodic interim results and maintaining honest, open, and frank communications.

Evaluators do their clients an injustice when they become preoccupied with the theoretical value of findings or lack confidence in the ability of the stakeholders to use data. University professors are often called upon to evaluate programs and sometimes they have difficulty remembering the audience for whom the report is being prepared. Reports that are heavy on statistics with little explanation about what they mean or the implications of these data for action are not always useful for program improvement. Clients and stakeholders need direct, clear prose that is devoid of education jargon and points the way for success.

Practical Procedures

To the extent possible, evaluation procedures should not be

disruptive to the ongoing activity of the program. Many sources of data collection exist within schools (see Figure 5.1) and can be used in both formative and summative evaluation designs. If possible, pilot the evaluation design and be sure that the design is appropriate given the time constraints.

Practical procedures relate not only to specific data collection strategies, but also such things as, "how contractual agreements with the client are reached; how data sources are chosen; which instruments are used and how they will be administered; how data and information are collected, recorded, stored, and retrieved; how data are analyzed; and how findings are reported" (p. 65).

The two most common errors relating to this standard include failing to weigh practicality against accuracy and disrupting program activities. This standard is particularly troublesome for studies where program implementation differs across sites. What may be practical for one site may be disruptive for another.

Teacher/student ratios	AFDC rate
Student/teacher portfolios	Graduation rate
Parent surveys	Repair bills
Attendance rates	Report cards
Accreditation reports	Teacher surveys
Free/reduced lunch figures	Library use
Promotion/retention rates	Book circulation
Guidance surveys	Title One report
Building volunteers	Dropout rate
Vandalism data	Site council attendance
Class size	AFDC rate
Test/assessment scores	Research studies
Student surveys	Suspension reports

Figure 5.1. Sources of data.

Political Viability

All programs are influenced by the political context in which they operate. Anticipating and understanding these various points of view are critical to the design and implementation of the evaluation. It is important for an evaluator to be cognizant of the politics of the situation prior to accepting the assignment. When political differences exist, it is helpful to provide period reports and identify, assess, and report different perspectives. There may be cases when the political environment is so volatile, conducting the evaluation would do more harm than good. In order to determine whether this is the case, the evaluator should meet with as many interest groups as possible prior to presenting a design or accepting a contract. The evaluator should "discontinue the evaluation if political issues create such an unfavorable situation that it appears the interests of all concerned will be best served by withdrawal" (p. 72).

Common errors related to this standard include relying solely on the clients' description of the program, failing to check for accuracy, and assuming that the program is uniformly implemented. If these errors are committed, the political wrangling about the study can completely discredit the work and cloud the participants' ability to use the data to improve the program.

Complete and Fair Assessment

The evaluation should address both the strengths and the weaknesses of the program. Evaluations that speak only to weaknesses can limit possibilities for program improvement. Some programs can be significantly improved by expanding the strengths of the program more easily than eliminating weaknesses. It is a good idea to send draft material to the district and request review and critical comments. Equally important is the necessity to report meaningful data that have been omitted from the evaluation.

Three common errors with this standard include giving the appearance that the evaluation is biased, failing to recognize both the formal and informal power structure of the system, and failing to provide caution and interim reports. In actuality, the final report should not come as a surprise. The data reported is based on what is happening in the district. While not often quantified, these data are known.

Program Documentation

This may sound like an easy standard to meet, but it is not. Two problems exist. First, the initial description of a program on paper is always different from what the program is when it is implemented. Frequently, the program conceptualization is so sketchy, that no one knows for sure what it is. Secondly, all programs "drift" over time. Two years after a program has been implemented, it is usually decidedly different from what it was in the beginning. This situation is exacerbated by changes in key personnel implementing or leading the program. The evaluator should ask both clients and stakeholders to describe the program. When the evaluator has written the program description he/she should ask for feedback regarding accuracy from everyone involved.

Common errors include failing to check for accuracy, writing either incomplete or superficial descriptions, and relying solely on paper descriptions of the program as the source.

Context Analysis

All programs operate within a context and they are influenced by that context. In order to interpret data accurately and effectively, the context in which the program operates must be described. This includes, "the geographic location of the program, its timing, the political and social climate surrounding it, competing activities in progress, the staff, and pertinent economic conditions" (p. 133). In describing the context, it is important to keep a log of unusual circumstances, events, or people operating within the environment that may influence the evaluation. It may also be useful to describe how the current program is the same as or differs from other similar programs where evaluations have been conducted.

Common errors with this standard include viewing the program's context too narrowly, concentrating too much on context, or ignoring potential influences. Reports should provide enough information about the context to enable the reader to understand the technical, social, political, organizational, and economic context of the program.

Justified Conclusions

All conclusions in an evaluation report should be explicitly justified. Without this information, stakeholders will tend to disregard conclusions. "Conclusions must be based on all the pertinent information collected, must incorporate results of sound analysis and logic, and must be accompanied by full information about how the evaluation was conducted" (p. 177). It is always prudent to advise the audience to be cautious in its interpretation of the conclusions and the number of conclusions discussed should be focused and limited.

Common errors include being too cautious, ignoring possible side effects, basing conclusions on insufficient or unsound information and failing to report the limitations of the study. The best evaluation reports are those that lead to actions to improve an effective program or terminate an ineffective one.

LIMITATIONS OF PROGRAM EVALUATION

No evaluation design could ever completely evaluate all aspects of a program. The program evaluator must decide what parts of the program are most important to evaluate. As depicted in Figure 5.2, program effectiveness is determined by an evaluation of a subset of the entire program. Poor alignment of a program evaluation design is represented by the figure on the right. In this example the evaluation design includes data collection procedures that do not relate directly to the program. The figure on the left shows a well-aligned evaluation. The information being collected relates directly to the program being implemented.

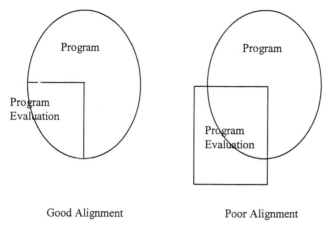

Figure 5.2. Program evaluation = measured program effectiveness.

TYPES OF PROGRAM EVALUATION DESIGNS

By defining program evaluation in a very broad sense, it could include three types of program evaluation designs: needs assessment, formative evaluation, and summative evaluation. Needs assessment answers the question, "What should we do?" Formative evaluation answers the question, "How is it working?" and summative evaluation answers the question, "Did it work?" The summative evaluation can become the needs assessment for the development of another program. In this way the program evaluation cycle can be seen as a cyclical process as depicted in Figure 5.3.

Some may argue that a needs assessment should not be considered a program evaluation design because these designs are most commonly attempting to discover "weaknesses or problem areas in the current situation which can eventually be remedied" (Herman, Morris, and Fitz-Gibbon, 1987, p. 15). Sometimes called an organizational review, questions posed by a needs assessment can include: "What needs attention? What should our program(s) try to accomplish? and Where are we failing?" (p. 16). Questions the evaluator might pose include the following:

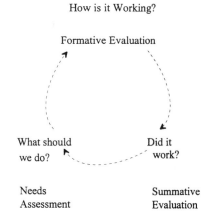

Figure 5.3. Type of program evaluation.

- What are the goals of the program?
- Do all stakeholders agree on the goals?
- Are the goals being met?
- Who are the clients involved?
- What do they need?
- What does staff need?
- What problems are being experienced?
- Is the organization addressing the problems perceived by clients and staff?
- To what extent?
- Where is the organization failing?
- Where are special programs needed?

Based on the outcome of the needs assessment, programs may be developed that would lead to both formative and summative evaluation designs.

A formative evaluation deals with program implementation. What is working and what needs to be changed? The questions guiding formative evaluation efforts include, "How can the program be improved? and How can it become more effective or efficient?" (p. 17). In designing a formative evaluation procedure the researcher may pose these questions:

- What is the purpose of the program, its goals and objectives?

- How would be program be described? What are its salient features?
- How is the program being implemented?
- What are the program components?
- Are they enabling the goals of the program to be achieved?
- How does the program need to be modified or adjusted?
- Is the program working for all clients, or are some clients benefiting more than others?
- What implementation problems have been encountered?
- What summative evaluation procedures should be identified?

Summative evaluation procedures are designed to answer the question, "Is the program working and if so, to what extent? Should the program be continued, modified, or eliminated? and What conclusion can be made about the effectiveness of the program?" (p. 17).

Many of the same questions asked in a formative evaluation should be addressed in a summative evaluation as well—such as, "Were the goals and objective of the program achieved? What are they? What are the most important features of the program?" Additional questions in a summative evaluation include, "Do the program benefits justify the costs involved? Are alternative programs available? Should they be considered? Is the program description now the same as it was at the beginning of the implementation? If not, how did it change and why? Should the program be continued?"

Given the huge demand for the limited resources available to educators, it is inconceivable that the educational community will continue to tolerate a situation where program evaluation is not a mandated part of the everyday life in schools. Today, it is common practice to identify a need; develop a new program; add it to existing programs; and after a few years, design still another new program without the benefit of a well-designed needs assessment, formative or summative evaluation. Educators traditionally avoid program evaluation designs. In the minds of many, program evaluation means personal evaluation. Without the benefit of program evaluation, we really don't know what is working or the extent to which it is working. Many school systems would define the K–12 reading program by a listing of the textbooks and supplementary materials used in the classroom, the number of minutes devoted to reading instruction, and the number and levels of certification of the teachers within the system. These systems would be hard pressed to define what the program is. For most, the program is what is assessed using state instruments. The success or failure of the program is defined by aggregate scores on these tests.

Based on a program description, an evaluator should be able to understand what happens in the program, the particular activities, processes, materials, and administrative arrangements included in the program, and be able to articulate a clear, consistent description of the aims and purposes of the program that are understood by all constituencies and guide program implementation and evaluation procedures. "Today, innovative programs as well as expansions of standard services can seldom be funded without some means of demonstrating that the costs of the service are justified by the improved state of the clientele" (Posavac and Carey, 1989, p. 4).

SOURCES OF RESISTANCE TO PROGRAM EVALUATION

In order to design an effective evaluation the evaluator will have to deal with resistance to conducting the evaluation during the early stages of the planning. Posavac and Carey (pp. 39–42) have identified nine sources of this resistance.

Expectations of a "Slam-Bang" Effect

People involved in designing a new program often set high expectations for themselves and expect to produce significant results in a very short period of time. They are often disappointed when the new program results don't demonstrate significant gains. The program evaluator can be helpful in assisting those involved in the program in setting reasonable objectives for improvement. This will decrease the possibility of the "slam-bang" effect and enable the program designers to look for long-term, gradual gains that may be more lasting.

Self-Styled Experts in Evaluation

Most of us have been involved in some form of program evaluation in our professional capacity. Sometimes, this involvement leads to the development of superficial or faulty evaluation designs. There is no recipe for an effective evaluation design that can be applied to every situation in exactly the same way. Every situation is different. While focus groups and structured interviews may be appropriate for one evaluation design, it doesn't mean these techniques should be used for all evaluation designs.

Fear That the Evaluations Will Inhibit Innovation

When an evaluation, either formative or summative, is underway, some participants may view this as a restriction to creative problem-solving. In these situations the need to maintain program identity must be balanced with the need to recognize opportunities for program improvement. All programs "drift" from the original conception of what the program was intended to be to what it actually becomes. It is important to document that drift and explain the benefits of the changes or time. Some recommend that it is not useful to evaluate a program that is just getting started because staff need to address the normal shift between what the program objectives were on paper versus what the program objectives become through implementation.

Fear That the Program Will Be Terminated

As the saying goes, "No news is good news." Some program participants fear that any information that is not confirming of a program's successful implementation will lead to termination of the program. During the first few years of a program, this is highly unlikely. Most new programs are funded and approved based on substantive data and a belief that the new program will solve a specific problem. Providing formative information about the implementation process should only serve to inform and enable program implementation improvement. Some have suggested that a new program may need four years of implementation before a determination can be made about continuation or termination. Providing program-effectiveness information during the course of those four years enhances the possibility that the program can be successfully modified and adjusted to meet its original goals.

Fear That the Information Will Be Abused

Teachers are traditionally fearful about the use of any evaluation of their performance. In a society that is becoming increasingly demanding that schools become more accountable, a tension is developing about the use of information. At the same time, with the movement to put more decision-making authority at the school level, schools are being asked to support their decisions with data. Data-driven decision making and avoidance of program evaluation data are not compatible. Once a faculty gets accustomed to using data, rather than "gut feelings" to make decisions, the tables will forever be turned. There is no evidence to suggest that decisions based solely on intuition are superior to decisions based on hard data.

Fear That Qualitative Understanding May Be Supplanted

Teachers feel "that their day-to-day observations are a valuable source of input both for improving the functioning of a program and for evaluating its effect. They may feel that the evaluators' questionnaires, complicated research designs, and statistical techniques are less sensitive than their own personal observation and evaluations" (p. 41). In actuality, both are important. Subjective input using qualitative means and objective input using scientific methods can both contribute effectively to program improvement.

Fear That the Evaluation Drains Program Resources

When evaluation strategies are built into a program implementation design, they are less likely to be intrusive or to drain the resources of the project. Evaluation designs, constructed "after the fact" are almost always more costly. In addition, it is difficult today to find a funding agency that does not require a description of the program evaluation design as part of the proposal. Unfortunately, these are sometimes quite superficial. The trend is toward the requirement that program evaluation designs be specifically developed prior to grant approval. Some agencies have started requiring their own evaluation component if they fund a project. These trends will only improve the data available for decision making and enable participants in the program implementation to become unafraid of data collection.

Fear of Losing Control of the Program

While common, this fear is probably groundless. Just because someone is collecting data regarding program effectiveness does not mean that the program evaluator has taken control of the project. Even though participants are not able to control the data that is being collected, the process works best when there is a good working relationship between the implementors of the program and the evaluator. A strong working relationship can greatly enhance the design of the evaluation and the value of the data being collected.

Fear That Evaluation Has Little Impact

This is becoming a less influential fear. With the dawn of the age of accountability, it is more important that the evaluation provide data that is most significant to determining program effectiveness. Knowing what part of the program to evaluate and how to do it is becoming a critical strategic decision. Ensuring that the data available is the appropriate data to make policy and funding decisions places a significant responsibility on the shoulders of the evaluator(s).

DEVELOPING PROGRAM EVALUATION DESIGNS

Since there is no universal program design that fits all occasions, constructing program evaluation procedures requires both artistry and skill. There are, however, some essential steps that should be completed. Generally, these steps fall into four phases: first, setting the parameters of the evaluation; second, selecting the evaluation methodology; third, collecting and analyzing the data; and finally, reporting the findings. The steps in each phase of the evaluation are outlined in Table 5.1.

Phase I—Setting the Parameters of the Evaluation

The first step in conducting an evaluation is to determine whether the evaluation is intended to be a formative or a summative assessment. The audience for a formative evaluation is involved in developing, managing, and implementing a program. In the case of a summative evaluation, the audience is

Table 5.2. Steps in creating a program evaluation design.

Phase I—Setting the Parameters of the Evaluation
1. Describe the purpose of the evaluation: is it a formative or summative design?
2. Find out about the program: economically, contextually, politically, and socially
3. Gain consensus regarding the program description
4. Determine the focus of the evaluation
5. Describe the evaluator's role

Phase II—Selecting Evaluation Methodology
1. Gain consensus regarding program description
2. Determine evaluation questions
3. Determine possible uses of study through stakeholder involvement
4. Design a plan for assessing program effectiveness or reviewing program implementation procedures
5. Design a time line for implementing the evaluation
6. Determine the cost of the study
7. Finalize contract

Phase III—Collecting and Analyzing Data
1. Implement evaluation design
2. Provide interim reports
3. Modify and adjust if necessary
4. Analyze data

Phase IV—Reporting the Findings
1. Determine procedures for communicating final report
2. Share draft copy with significant stakeholders
3. Revise, edit, finalize
4. Submit report

Source: Adapted from Herman, Morris, and Fitz-Gibbon, 1987.

more likely to be policy makers, funding agencies, and various interested publics. The role of the evaluator will be different for each type. In formative evaluation designs, the evaluator tends to work in a collaborative, interactive way with those implementing the program. In a summative evaluation design, the evaluator may become more independent from those implementing the program serving as data providers. Data collection procedures for both types typically involve qualitative and quantitative measures, although summative designs may rely more heavily on quantitative data and formative designs may include more qualitative techniques, such as focus groups and structured and open-ended interviews. In formative designs, the mechanisms for sharing findings may be more informal and rely on both interim reports and oral presentations, whereas summative evaluation findings almost always include a final written report.

In the course of developing an accurate, agreed-upon description of the program, the evaluator will have to read a variety of materials and talk with a number of people representing different constituencies. A clear description of the program must be established prior to the development of the evaluation design. Asking various groups to complete the Program Evaluation Worksheet in Table 5.2 may enable the researcher to gain a more accurate understanding of how various groups view the program. By asking several individuals or groups to complete this worksheet, the evaluator can get an assessment of the amount of divergence that exists in program description, purpose, and reason for the evaluation. The amount of time needed to complete Phase I will depend upon the information received using this worksheet. Phase II should not be initiated until the role of the evaluator has been clearly defined within the scope and focus of the evaluation design.

Phase II—Selecting Evaluation Methodology

Selecting evaluation methodology should be a collaborative effort between the evaluator and the client(s) with due consideration for the parameters set by the context, budget, and environment in which the study will be conducted. Before the

Table 5.2. Program evaluation worksheet.

Program Name:
Program Description (site, age, population served, staff, budget, etc.):
Program Purpose(s):
Reason for Evaluation:
Stakeholders—Who is interested?
Questions that need to be answered:
Data available:
Data needed:

design is set, there should be consensus about the operation of the program, including the goals and objectives of the program, the primary activities and organizational arrangements, the roles and responsibilities of the projects staff, and the outcomes the program is designed to achieve (Herman, Morris, and Fitz-Gibbon, 1987, p. 58).

At this point, the evaluator should be in a position to construct the program evaluation questions that will serve to guide the development of the design. These questions should be specific. It is a good idea to check your questions to be sure that they address the primary concerns of the staff. Listen to the feedback, and if necessary, revise the questions.

In order to be sure that the work can progress smoothly, it is a good idea to develop a program monitoring plan. This plan should include a listing of the activity/components to be examined, data collection techniques or instruments selected, dates of data collection, sites to be examined and subjects participating in the study. This plan serves as an outline of the work. If the study involves a control group, it is helpful to construct a second program monitoring plan for that group.

One of the most difficult tasks to be completed is determining the cost of the study. Clients should be provided with as much detail in determining the costs as possible. Generally, these costs fall into four areas: evaluation staff, materials, and supplies; travel; consultant fees (if appropriate); and special costs. If more than one evaluator is working on the project, the daily stipend and fringe benefits may differ, depending on the responsibilities and expertise of the evaluators. This category should also include the costs for secretarial and support services, photocopying, telephones, office equipment, and utilities. Travel expense will depend on the number of people traveling, the number of trips and the length of each site visit. These costs can vary dramatically, depending on whether the evaluators are traveling by air or by car. If the project calls for the use of consultants other than the evaluation staff, their stipend and expenses should be itemized separately. The duties of these consultants should be explicitly described in the evaluation design. Finally, if there are special costs for printing or the purchase and mailing of surveys, these should be itemized. If the final budget is prohibitive, there are measures that can be taken to reduce it. Herman, Morris, and Fitz-Gibbon suggest nine (1987, p. 72):

1. Rather than interviewing the entire staff, it may be possible to sample some staff for personal interviews and use a survey to collect data from the rest of the staff.
2. Train and employ junior staff to complete some of the data collection and analysis.
3. It may be possible to utilize some of the staff employed in the project to assist in data collection without jeopardizing the reliability and validity of the data.
4. Purchasing assessment instruments may be more economical than developing new ones.
5. Utilizing procedures and staff you have previously trained may cut down on planning time.

6. Maximizing on-site time to collect as much information as possible from a variety of groups can cut down on the number of site visits required.
7. Reducing the scope of the work addresses those factors of greatest importance.
8. Utilizing machine-scored instruments.
9. Utilizing information that is already routinely collected as part of the evaluation design.

The final evaluation design agreement should be presented to the client(s) in writing and be formally approved. This may include the public action of a formal board. The evaluator should also request approval of the plan in writing. As the work evolves, there may need to be changes in the design. All of these changes should be documented in writing and appropriate approvals provided. These actions serve as a protection to both the evaluator and the client. Without them, the scope of the work tends to be expanded, and both the client(s) and the evaluator may become dissatisfied with the way the project evolves.

Collecting and Analyzing Data

Someone once said, "First you plan the work, and then you work the plan!" Collecting and analyzing the data is an example of working the plan. During this phase of the process, the attention of the evaluator is on seeing that the work gets done in a timely and effective manner. Particular attention should be paid to the time lines established in the plan. If specific steps in the process are not completed on time, it throws off the entire sequence of events. Depending on the length of the project and the political factors influencing the outcome, the evaluator may want to develop interim reports. If it is necessary to modify or adjust the initial plan, it should be done with the full knowledge of why the adjustments are necessary and how these adjustments will impact the overall design of the project.

Great care must be taken during this phase of the project to be sure that the data being collected are accurate. It may be necessary to double check both the accuracy of the instruments being used and their match with the program objectives. As the analysis begins to take shape, additional questions may arise. At this point, it will be necessary for the evaluator to determine whether additional information is required and, if it is, how it will be obtained.

Reporting the Findings

The purpose of any evaluation design is action. Action leading to expansion, modification, or termination of a program. The format and substance of the reporting procedures should facilitate that action. An evaluation report that is received and placed on a shelf, with no action taken, is reflective of a poorly conceptualized, executed, and reported design. Whatever communication design selected for the final report, it would help facilitate action. Whether it is a formative or summative evaluation design, it is important to meet with program staff to relay the findings, even though they are probably not the primary clients. Often, the clients are in a funding or policy-setting role, such as a board of education. The board may or may not know what to do with the findings. If the evaluator has met with the staff of the project and made them aware of the findings and recommendations, the staff is in a better position to assist the board in determining the appropriate "next steps" to take.

Time is short and there is much to do. Effective program evaluation can help reduce the time required to modify and adjust educational practice in order to maximize student achievement. Program "evaluation is a diligent investigation of a program's characteristics and merits. Its purpose is to provide information on the effectiveness of projects so as to optimize the outcomes, efficiency, and quality" (Fink, 1995, p. 2). As we move toward developing curriculum and assessment strategies to equip students with the skills and knowledge they will need in the 21st century, program evaluation designs will play a significant, integral part.

APPENDIX

The Program Evaluation Standards

Noncopyrighted material—Reproduction and dissemination are encouraged. Developed by The Joint Committee on Standards for Educational Evaluation Published by Sage Publications, Inc.

Utility Standards

1. Stakeholder identification—Persons involved in or affected by the evaluation should be identified so that their needs can be addressed.
2. Evaluator credibility—The persons conducting the evaluation should be both trustworthy and competent to perform the evaluation so that the evaluation findings achieve maximum credibility and acceptance.
3. Information scope and selection—Information collected should be broadly selected to address pertinent questions about the program and be responsive to the needs and interests of clients and other specified stakeholders.
4. Values identification—The perspectives, procedures, and rationale used to interpret the findings should be carefully described so that the bases for value judgments are clear.
5. Report clarity—Evaluation reports should clearly describe the program being evaluated including its context, and the purposes, procedures, and findings of the evaluation so that essential information is provided and easily understood.
6. Report timeliness and dissemination—Significant interim finds and evaluation reports should be disseminated to intended users so that they can be used in a timely fashion.
7. Evaluation impact—Evaluations should be planned, conducted, and reported in ways that encourage follow-through by stakeholders so that the likelihood that the evaluation will be used is increased.

Feasibility

1. Practical procedures—The evaluation procedures should be practical to keep disruption to a minimum while needed information is obtained.
2. Political viability—The evaluation should be planned and conducted with anticipation of the different positions of various interest groups so that their cooperation may be obtained, and so that possible attempts by any of these groups to curtail evaluation operations or to bias or misapply the results can be averted or counteracted.
3. Cost effectiveness—The evaluation should be efficient and produce information of sufficient value so that the resources expended can be justified.

Propriety

1. Service orientation—Evaluations should be designed to assist organizations to address and effectively serve the needs of the full range of targeted participants.
2. Formal agreements—Obligations of the formal parties to an evaluation (when it is to be done, how, by whom, when) should be agreed to in writing so that these parties are obligated to adhere to all conditions of the agreement or to formally renegotiate it.
3. Rights of human subjects—Evaluations should respect human dignity and worth in their interactions with other persons associated with an evaluation so that participants are not threatened or harmed.
4. Human interactions—Evaluators should respect human dignity and worth in their interactions with other persons associated with an evaluation so that participants are not threatened or harmed.
5. Complete and fair assessment—The evaluation should be complete and fair in its examination and recording or strengths and weaknesses of the program being evaluated

so that strengths can be built upon and problem areas addressed.
6. Disclosure of findings—The formal parties to an evaluation should ensure that the full set of evaluation findings along with pertinent limitations are made accessible to the persons affected by the evaluation and any others with expressed legal rights to receive the results.
7. Conflict of interest—Conflict of interest should be dealt with openly and honestly so that it does not compromise the evaluation process and results.
8. Fiscal responsibility—The evaluator's allocation and expenditure of resources should reflect sound accountability procedures and otherwise be prudent and ethically responsible so that expenditures are accounted for and appropriate.

Accuracy

1. Program documentation—The program being evaluated should be described and documented clearly and accurately so that the program is clearly identified.
2. Context analysis—The context in which the program exists should be examined in enough detail so that its likely influences on the program can be identified.
3. Described purposes and procedures—The purposes and procedures of the evaluation should be monitored and described in enough detail so that they can be identified and assessed.
4. Defensible information sources—The sources of information used in a program evaluation should be described in enough detail so that the adequacy of the information can be assessed.
5. Valid information—The information gathering procedures should be chosen or developed and then implemented so that they will assure that the information obtained is sufficiently reliable for the intended use.
6. Systematic information—The information collected, pro-

cessed, and reported in an evaluation should be systematically reviewed and any errors found should be corrected.
7. Analysis of quantitative information—Quantitative information in an evaluation should be appropriately and systematically analyzed so that evaluation questions are effectively answered.
8. Analysis of qualitative information—Qualitative information in an evaluation should be appropriately and systematically analyzed so that evaluation questions are effectively answered.
9. Justified conclusions—The conclusions reached in an evaluation should be explicitly justified so that stakeholders can assess them.
10. Impartial reporting—Reporting procedures should guard against distortions caused by personal feelings and biases of any party to the evaluation so that evaluation reports fairly reflect the evaluation findings.
11. Meta-evaluation—The evaluation itself should be formatively and summatively evaluated against these and other pertinent standards so that its conduct is appropriately guided and on completion, stakeholders can closely examine its strengths and weaknesses.

REFERENCES

Fink, Arlene. (1995). *Evaluation for education and psychology.* Thousand Oaks, California: Sage Publications, Inc.

Herman, Joan, Morris, Lynn Lyons, & Fitz-Gibbon, Carol Tylor. (1987). *Evaluator's handbook.* Newbury Park, California: Sage Publications.

Posavac, Emil J. & Carey, Raymond G. (1989). *Program evaluation: Methods and case studies.* Englewood Cliffs, New Jersey: Prentice Hall.

The Joint Committee on Standards for Educational Evaluation. (1994). *The program evaluation standards: How to assess evaluations of educational programs, 2nd ed.* Washington, D.C.: American Psychological Association.

CHAPTER 6

Toward Continuous Curricular Improvement

CURRICULUM anchored temporally to meeting emergent social needs soon becomes outdated since such needs are rarely stable. Even if stability is assumed, when needs are met, they cease to be so because new needs soon emerge. The rising and falling tide for engineers amply illustrates this point. At one point, U.S. engineering schools were not producing enough engineering graduates. At other times, engineering graduates can't find jobs and there is a glut (see Magner, 1995). The educational sector most at risk of becoming a passing fad is vocational education, especially in the fast changing sector of technology.

There is an array of factors that curriculum developers must balance in creating a multiplicity of work plans. Let us examine some of the influences that impinge on curriculum change and development in the United States. These are unique because of the highly decentralized nature of political influence in the construction of curriculum.

NATIONAL PRIORITIES

Despite the rancor kicked up by the Goals 2000 effort, the public schools have been responsive to national needs in the past (see Howe, 1991). One thinks of the Russian Sputnik that triggered the National Defense Education Act of 1958, which was the first big push for schools to become involved in the Cold

War. Centered first on science, math, and foreign language acquisition, the NDEA was later expanded to include more curricular subject areas.

The Elementary and Secondary Education Act of 1965 was the next large federal effort to change the nation's schools. ESEA was divided into a number of titles. Title I became the big push on poverty, particularly in attacking problems with basic skills such as reading. Title II was devoted to school libraries, textbooks, and other media. Title III pertained to promoting local educational innovations. Title IV sponsored research and development centers. Title V related to improving state departments of education. Curriculum change was part of the effort to improve education and schools within ESEA. Its major flaw was that "the social and economic system that had created poverty and allowed it to continue was not considered the problem; the problem was the culture of the poor" (Spring, 1986, p. 309).

ESEA began the great war on poverty that was, in reality, a war on the culture of the poor. In the days of Lyndon Johnson, the schools were seen as the bridge by which the poor could avail themselves of the ladder toward the middle class. The idea that the school was a barrier for the poor to ever reach the middle class was simply not ever considered. Education had become depoliticized (Apple, 1982, p. 20). What was simply needed was more efficient means to make schools as understood run better, faster, and cheaper. This particularly virulent form of conceptualizing school-related problems has a long history in education, particularly in the preparation of school administrators (see Callahan, 1962).

Now, after several decades of federal and state funding centered on the idea of refashioning schools to become more efficient, the move is to abandon public schools altogether to market economics with charter schools (see Fine, 1994) and voucher plans (Chubb and Moe, 1990). A silent consensus among some reformers has been forged that the public schools cannot be saved. There is too much interference from legislative mandates, teacher unions, and government bureaucrats (Finn, 1991). Once again, the corporate culture ideology of "freeing the schools" from the tentacles of mindless legislation, union interests that produce mediocrity instead of excellence, can only be thwarted by taking the schools to the people directly and letting them vote with their vouchers.

These same antidotes ignore the problem of the class structure and its stake in maintaining socioeconomic inequity and hegemony (Giroux, 1981, p. 92). The same ideology in America 2000 prevails that poverty exists because the poor have an inherently inferior culture or are genetically stupid and so deserve their fate (see Herrnstein and Murray, 1994). No national effort in education has ever attacked the root cause of poverty, which is the economic system itself.

A Nation at Risk, the clarion call of the Reagan Administration, was steeped in the rhetoric of efficiency and Cold War metaphors. It warned that America would lose the war globally if its schools did not become more competitive. It also spawned the idea of "at-risk" children. Singer (1985) has indicated that *A Nation at Risk* conformed to the following predetermined goals (pp. 355–356):

1. That the educational system represented a great mediocrity
2. That America was falling behind militarily and industrially in the world
3. That the "new basics" must replace the old ones
4. That a series of cures be prescribed, which ultimately meant more bureaucracy, more tests and more basics in schools
5. That local and state agencies should ante up more money to pay for these changes

A Nation at Risk ushered in no new or radical curricular reforms in American education. The antidotes were more of the same that was already underway.

STATE INITIATIVES

The various states have initiated a broad band of initiatives in attempting to "fix" schools. Virtually all fifty states have sponsored a variety of curricular changes, most related to adding on to graduation requirements (see Roberts and Murray, 1995). The efforts of the states have mirrored those at the federal level. Nearly all of the assumptions behind federal intervention have been accepted at the state level as well.

The history of the common school movement in America was the product of state reforms begun by Horace Mann in Massachusetts and Henry Barnard in Connecticut in the 1830s and 1840s (see Kaestle, 1983). The influence of reform has been carried forward at the state level by the introduction of "high-risk" or "high-stakes" performance assessment (Guskey, 1994).

FOUNDATION FUNDING

Private foundations have sponsored a variety of change thrusts in American education. Carnegie, Rockefeller, Danforth, and others have aligned themselves with various types of change programs and agendas. To date, none have enjoyed a major breakthrough in improving opportunities for all students or in reducing class divisions based on wealth in American society.

PROFESSIONAL BASED INITIATIVES

Some change efforts are almost exclusively confined to within professional channels, eventually emerging into the political mainstream. One thinks here of the "behavioral objectives" movement of the 1960s; the accountability movement, which still exists (Wagner, 1989); the Madeline Hunter effective teaching concepts; and lately, the OBE (outcomes-based education) idea, which has appealed to administrators (Dlugosh, Walter, Anderson, and Simmons, 1995) and brought down the wrath of the religious right (Manatt, 1995), which quashed or diluted much of its intent.

LOCAL BOARD INITIATIVES

Local school boards still exercise the power to initiate change in American curricula. There are notable examples of widespread efforts to alter curriculum. One is the Chicago Mastery Learning curriculum, a K–8 program implemented in 1979 and abandoned some years later (see Gibboney, 1994, pp. 104–126). While the power to initiate curricular change has historically been lodged with local boards of education, only a few have actually used it to engage in true broad-based innovation. Boards remain inherently conservative in their orientation to curricular issues because they represent civic elites, that is, bankers, industrialists, and heads of local businesses (Spring, 1986, p. 225) who have a vested interest in current political relationships, especially the economic status quo.

CURRICULUM DEVELOPMENT AS POLITICAL CONSENSUS

While there is discussion in academic texts regarding the reference points for curriculum development involving the needs of children and their developmental maturation, the creation of curriculum as work plans is almost exclusively political and external to schools.

Curriculum "cores" are created that involve varying merged political consensus at the state level in the form of "frameworks" or guidelines. These take the shape of required courses of study, Carnegie Units accrued in various curricular content areas (history, math, science, foreign language) or broad-based goals or expectations, which may or may not be located in a curricular discipline.

Once a political consensus has been achieved, "core" frameworks are then translated into logical derivatives based on the idea of simple to complex sub-units/skills or knowledges or from a perspective of primitive psychological notions of learning possibilities, most often Skinnerian in form (a variation of simple to complex).

So little is known about the psychology of learning beyond a few principles from behaviorism or a view from linguistic studies that any discussion regarding the continuous improvement of curriculum starts outside what learners may bring to learning situations as places to define curriculum. We are not embracing this perspective but merely describing contemporary practice.

Public schools are quasi state agencies and as such, must function politically within bands of acceptance by the public who elect legislators and members of boards of education to oversee curricular matters. Few legislatures or boards begin a discussion about curriculum from psychology. They almost always begin by considering the question "What is worth learning/teaching?" They assume that the "educators" are the ones who will arrange or sequence a curriculum appropriately so that it can be learned/taught to age-specific clientele in schools.

The idea of the continuous improvement of curriculum as envisioned in contemporary practice refers to the learning of an externally defined curriculum subdivided into simple to complex tasks by students and the creation of "more" curriculum once this has occurred. It also refers to the idea that "new knowledge" must be infused into the work plan as it becomes known and approved by state agency officials. "New knowledge" usually refers to the passing of history and events (teaching about one more war or past events such as the removal of the Berlin Wall) or updated information created by research (as in science) or in technology (newer computers).

The lines of political consensus do not always parallel what research clearly indicates, particularly if such research contravenes social practices anchored in religious perspectives. For example, there is substantive medical research that indicates that moderate alcohol consumption (one to two drinks per day) conclusively reduces the risk of heart attack and lowers the risk of coronary artery disease. Heart disease is the number one killer of Americans (Chase, 1995). Few legislatures or boards of education have suggested incorporating this definitive research in any of the curriculum for which it may be applicable (science, home economics, health) and not many teetotalers are expected to invite Jim Beam home for dinner either.

A similar situation has occurred with knowledge about contraception. While much is known about the prevention of pregnancy, the only remedy acceptable to a broad band of Americans to be included in school curricula is total abstinence of any sexual activity at all, a solution that does not appeal to a large part of the teenage population (Teen Trends, 1995, p. 4). The use of research as a base upon which to develop curriculum is clearly limited to what can find broad-based political appeal to voters who elect politicians and that does not contradict other perspectives they may hold.

When the American Association for the Advancement of Science (1989) declared that, "Most Americans are not scientifically literate" (p. 13) and trumpeted that, "There are no valid reasons—intellectual, social, or economic—why the United States cannot transform its schools to make scientific literacy possible for all students," they ignored the political reality that confuses the theory of evolution with "scientific creationism," which is neither a theory nor science. Any board of education or state legislature that refuses to acknowledge that as a fundamental fact cannot be expected to embrace scientific literacy. There are political reasons why Americans are not scientifically literate, which often escape the public because the public schools are responsive to public attitudes that are unscientific.

So there are clearly limitations upon which the continuous improvement of curriculum can occur in public schools. That improvement must conform to prevalent social mores and opinions that are often contradictory to what research has indicated is a fact. The continuous improvement of curriculum means that there is some systematic and structural response built into schools and school systems that works as a kind of standard operating procedure to update and change the curriculum. The sources of these potential changes are described briefly.

INTERNATIONAL STANDARDS AND TESTS

Emerging international standards in education (see Chapter 3) are beginning to set the global definition of a world-class curriculum. Such standards will not be defined by any local school board or state legislature. World-class standards for curriculum are those embraced by a consensus of world nations. The resistance to the emerging movement to attain broad political consensus about national U.S. educational goals will ultimately mean that international tests will determine the curriculum instead of the curriculum focusing the test. The inability of the American public to come to terms with an approach where educational goals are defined in any other way than locally is a major barrier to international educational comparisons for Americans because it means that many U.S. students may be tested on a curriculum for which they have not been adequately taught (see Westbury, 1992).

NATIONAL FAILURES TO IMPROVE TEST SCORES

While the discourse regarding a national curriculum have faltered, the movement toward some sort of national exam continues. That exam or test assumes that some sort of common curriculum is in place by which to make comparisons among state, city, or district possible. Without a public consensus on national educational goals, a sub rosa national curriculum exists in the content that is tested. In such comparisons, the dominance of SES will continue to provide the largest share of the explanation of the variance among scores because tests are not culturally fair nor free (see House and Haug, 1995).

If accountability mechanisms accompany the national exam, it can be expected that there will be "drift" toward alignment to the test. The national test may drive the creation of a national curriculum far more than a reasoned debate on the front end regarding what is worth teaching or knowing.

BATTLES OVER PRIVATIZATION AND VOUCHERS

The failures of the public schools, particularly in urban areas, have provided the grist to turn schools over to for-profit corporations to run them. Few such efforts have resulted in significant achievement gains, though some show improved financial management practices. Unless such schools are magnet schools, they function within state-approved curricular guidelines so they become merely the means to show efficiency improvements within the existing curriculum framework.

The future may bring the opportunity to experiment with different curricula. That would open up the curriculum development process to individual schools. At least theoretically, the empowerment of local schools implies that a very different curriculum could be constructed.

CREATING A SYSTEMIC RESPONSE

Forming a Policy Advisory Committee

Local school boards should see that some sort of permanent curriculum council exists in their school systems by which external mandates can be considered (whether by tests or curriculum content/standards). Such bodies can be expected to be policy advisory committees to the superintendent and the board that would suggest changes in the boards' policies. This group would be expected to debate the pros and cons of alterations in student expectations, courses, curricular sequencing, graduation requirements, textbooks adoptions or challenges, and local responsiveness. The group might include parents, representatives of the business community, citizens, and/or students. The group is not a technical working body, but a body politic to consider policy issues.

Creating a Strategic Plan

Once policies have been formulated, it will be necessary to incorporate them into a generic planning document that communicates and translates them into specific actions over a definitive time period. A strategic plan is a rational response to meet the need for the school system to focus itself on critically important tasks and actions.

Part of the creation of a strategic plan will be using selected visions of the future. Various futurists have proffered different perspectives (see Naisbitt, 1982). For example, Cetron (1985) has pointed out that one inevitable trend in the larger society is the move from an industrial society to one of an information-age society. The requisite need for educated people shifts from a focus on simple disciplinary skills to interdisciplinary programs with varied program options (p. 135). Cetron also notes that the requirements for a unicultural focus gives way to a multicultural focus. An information society requires higher-order skills and the use of the computer not as a vocational tool, but as essential as reading itself as a universal learning tool.

"Visioning" is a process of scanning the horizons of the future so as to avoid perpetuating the obsolescence of the past. As such, it must utilize a data base that is centered on the future. A strategic plan is the institutional embodiment of an effort to rationally examine the future and use the analysis to ensure that children are prepared for their times instead of their parents' times. Visioning is not a look back to some nostalgic past that never really existed (the "good old days"). Rather, it is a hard look into the unknown, the uncomfortable, and even the unthinkable. Because of system politics, the visioning process will always be subject to compromises. There are no "pure" views possible outside of the present value-based structure.

Creating Operational Indicators and Action Plans

A strategic plan must be translated into tangible yearly targets and related to the annual budgeting process. This process reduces a multi-year focus to a year-by-year focus. These, in turn, must be shifted into an organizational structure that is filled with people occupying offices that have relationships to one another (see Nadler, Gerstein, and Shaw, 1992).

The translation process occurs several ways. Multi-year targets divided into vertical organizational relationships can be impacted in the development of job descriptions and line/staff charts which provide the lateral (horizontal) scope of work to be done (see Galbraith and Nathanson, 1978).

Annual job targets can be translated using derivatives of the "management by objectives" approach. These can be broken down into the "results" expected and related to the work of individuals in the organization (Watson, 1981).

The steps in developing and selecting educational indicators have been outlined by Blank (1993, p. 67):

1. Develop a conceptual framework based on research results and the interests of policy makers and educators.
2. Obtain commitment and cooperation of leaders.

3. Involve policy makers, educators, researchers, and data managers in selecting priority indicators.
4. Select a limited number of indicators and minimize complexity of reporting.
5. Organize a cooperative data system that is based on determining appropriate methods of collection and the use of standards for producing comparable data.
6. Report comparative data on indicators to users and stakeholders.

Evaluating Performance and Results

Formal evaluations of progress toward annual objectives should be contained in year-end administrative reports (see Chapter 5). These should be accompanied by candid revelations about what worked and what didn't. Specific recommendations should accompany the annual report. Such reports should be developed and made public long before the next cycle of budgetary objectives is established. The idea is to establish a "curriculum driven" budget (Greene, 1995, p. 222).

Reestablish Annual Targets Based on Feedback

Each budgeting cycle is formulated around tangible achievement of curriculum centered priorities. Annual updates occur within the selected organizational structure and may be reflected again in the yearly management by objectives format (Odiorne, 1965).

Reappraise Strategic Plan

Every three to five years, the overall strategic plan should be seriously reexamined for major revisions. Minor revisions can occur annually. Questions that should be answered are: How do we know we are still on the right track? Are there indicators we have selected an inappropriate vision? Have we ignored them? Are the means still appropriate to attain the selected ends? Is organizational performance adequate? Will we be able to reach our vision? If not, why? What adjustments appear to be required to improve the capability of the school or district to reach its objectives?

The reappraisal of the strategic plan is centered on the real world dilemma that school curriculum must be delivered within a school system, that is, an organization. Sometimes curriculum people appear to believe that curricular issues can be debated while ignoring the practicality that organizations place demands on services or processes incorporated into its design. Organizations possess rules, and are ordered and not "natural" places. No curriculum is context-free. All school curriculum must be implemented within a structure. Discussions about structures are therefore relevant to curriculum debates.

The requirements for accountability and control often run contrary to curricular ideas of letting classroom teachers ignore the realities of system life, particularly if students are required to take state tests of some sort. Improving student achievement (the learning that is tested) means maintaining both vertical and lateral focus and connectivity with alignment. In such cases, teacher selection is reduced to creating effective means within a framework, as opposed to ignoring it.

Much of the change literature that urges individual school unit autonomy on all matters ignores the necessity of forging vertical congruence on a common curriculum in order to permit secondary school students to succeed when their performance is largely a measure of the alignment of their high school curriculum to that of the elementary and middle schools. Tests assess cumulative and prior learning. Poor secondary school test performance can be indicative of the lack of connectivity to tested content that is dependent upon prior cumulative learning. In this case, while the test is administered at the secondary level, it is more a measure of the elementary and middle school curricula than the high school. If elementary and middle schools are allowed to choose their own curriculum that is not tested, they succeed at the expense of the high school, a condition known as sub-optimization (English, 1987, pp. 289–290). If each school is a school system unto itself, then a test should be devised that measures only their contribution. Drawing curricular lines around such tests is nearly impossible. The more a test is administered later in a student's career, the more difficult it is to pinpoint the source of the knowledge learned to successfully pass it. The curriculum may be compartmentalized, but learning transcends all such boundaries.

Engaging in continuous curriculum improvement means not taking any chances that a single response is adequate. It means placing the entire curriculum design and delivery process into a planned, internal, and systemic set of organizational relationships. It means using curriculum to shape the budgeting process, not the other way around. And it means bold visioning to look past the status quo into an ambiguous future. Only the past promises certainty, and that is an illusion.

Curriculum improvement always exists in the caldron of practical politics that compromises everything, including the whole truth about anything. For this reason, curriculum development is always an incomplete business. It is an activity not for the faint of heart, for nothing is more fought over than the values of life held dear.

REFERENCES

American Association for the Advancement of Science. (1989). *Science for all Americans.* Washington, D.C.: Author.

Apple, M. W. (1982). *Education and power.* Boston: Routledge & Kegan Paul.

Blank, R. K. (1993, Spring). Developing a system of education indicators: Selecting, implementing, and reporting indicators. *Educational Evaluation and Policy Analysis,* 15(1): 65–80.

Callahan, R. E. (1962). *Education and the cult of efficiency.* Chicago: University of Chicago Press.

Cetron, M. (1985). *Schools of the future.* New York: McGraw-Hill.

Chase, M. (1995, July 3). Beneficial drinking: After abstinence, before tying one on in Health Journal. *Wall Street Journal.* B-1.

Chubb, J. and Moe, T. (1990). *Politics, markets, and America's schools.* Washington D.C.: The Brookings Institution.

Dlugosh, L. L., Walter, L. J., Anderson, T. and Simmons, S. (1995, April). OBE: Why are school leaders attracted to its call? *International Journal of Educational Reform,* 4(2): 178–183.

English, F. W. (1987) Curriculum management. Springfield, Illinois: Charles C Thomas Publisher.

Fine, M. (1994). Chartering urban school reform. In Fine, M. (Ed.). *Chartering urban school reform.* New York: Teachers College Press.

Finn, C. E., Jr. (1991). *We must take charge.* New York: The Free Press.
Galbraith, J. R. and Nathanson, D. A. (1978). *Strategy implementation: The role of structure and process.* St. Paul, MN: West Publishing Company.
Gibboney, R. A. (1994). *The stone trumpet.* Albany, New York: SUNY Press.
Giroux, H. A. (1981). *Ideology, culture, and the process of schooling.* Philadelphia, PA: Temple University Press.
Greene, S. (1995). Standard five: The productivity standard. In L. Frase, F. W. English, and W. K. Poston, Jr. (Eds.). *The curriculum management audit.* Lancaster, PA: Technomic Publishing Co., Inc., pp. 213–229.
Guskey, T. R. (1994). *High stakes performance assessment.* Thousand Oaks, CA: Corwin Press.
Herrnstein R. J. and Murray, C. (1994). *The bell curve: Intelligence and class structure in American life.* New York: The Free Press.
House, E. R. and Haug, C. (1995, Summer). Riding the bell curve: A review. *Educational Evaluation and Policy Analysis,* 17(2): 263–272.
Howe, H. (1991, November). A bumpy ride on four trains. *Phi Delta Kappan.* 192–203.
Kaestle, C. F. (1983). *Pillars of the republic.* New York: Hill and Wang.
Magner, D. K. (1995, June 30). Excess production of Ph.D.'s found in engineering and some sciences. *The Chronicle of Higher Education,* XLI(42): A-16.
Mannatt, R. P. (1995). *When right is wrong.* Lancaster, PA: Technomic Publishing Co., Inc.
Nadler, D. A., Gerstein, M. C. and Shaw, R. B. (1992). *Organizational architecture.* San Francisco: Jossey-Bass Publishers.
Naisbitt, J. (1982). *Megatrends.* New York: Warner Books.
Odiorne, G. S. (1965). *Management by objectives.* New York: Pitman Publishing Company.
Roberts, R. R. and Murray, K. T. (1995, April). State school reform legislation and related litigation. *International Journal of Educational Reform,* 4(2): 153–161.
Singer, I. (1985). What's the real point of *A Nation at Risk?* In Gross, B. and Gross, R. (Eds.). *The great school debate.* New York: Simon and Schuster, pp. 354–357.
Spring, J. (1986). *The American school 1642–1985.* New York: Longman.
Teen Trends. (1995, June 21). *Education Week,* 14(39): 4.
Wagner, R .B. (1989). *Accountability in education.* New York: Routledge.
Watson, C. E. (1981). *Results-oriented managing.* Reading, MA: Addison-Wesley Publishing Co.
Wheeler, Patricia & Haertel, Geneva. (1993). *Resource handbook on performance assessment and measurement: A tool for students, practitioners, and policy makers.* Livermore, California: The Owl Press.
Westbury, I. (1992, June–July). Comparing American and Japanese achievement: Is the United States really a low achiever? *Educational Researcher,* 21(5): 18–24.

CHAPTER 7

A Keyhole Peek into the 21st Century

THERE is much talk about the 21st century. American presidents have waxed and waned about being number one in education in the world, yet there is precious little in the way of substantive changes to ensure that political rhetoric is translated into a reality. The outline of the 21st century has been clearly sketched out. It will be called the "Asian Century" because of the rising economic and political power of Japan and the four Asia tigers of Singapore, Hong Kong, Taiwan, and South Korea. In the early 1960s, South Korea could not produce a simple battery. In 1995, it is a leading manufacturer of sophisticated computer chips (Biers, 1995, p. A19). In Taiwan, the Hsinchu Science Park rivals California's Silicon Valley (Biers, 1995, p. A19).

China has already become an exporting colossus, creating a $30 billion trade surplus that might surpass Japan's. China now makes two-thirds of the world's shoes and half of the world's toys. It makes most of the world's bicycles, laps, power tools, and sweaters. In electronics alone, China's production of fax machines expanded thirty-fold in the first nine months of 1995 (Kahn, 1995, p. A1). Already the financial pages of the *Wall Street Journal* herald the bulging muscles of Asian prosperity and power. The largest public company in the world is NTT of Japan, as well as the largest insurance company in the world (Nippon Life) ("The World's 100 Largest Banks," 1995, p. R33). When the 100 largest banks in the world are ranked, the first eight are all Japanese (p. R33). Citicorp is the highest ranked financial institution from America at a mere twenty-eighth. The United States is no longer the wealthiest nation in the world. A study by the World Bank of nations' per capita wealth shows America is only twelfth, behind Japan, Sweden, Australia, Canada, Norway, Denmark, and the United Arab Emirates (Wessel, 1995, p. A2).

A recent study by the Paris, France–based Organization for Economic Cooperation and Development showed that the widest gap in income between rich and poor exists in the United States out of the world's twenty-five richest countries (Bradsher, 1995, p. C2). The United States leads the world in the number of missing fathers in homes. Twenty-five percent of one-parent households with dependent children are without a father in America ("The Family," 1995, p. 26). In America, one of three black men in their twenties are in prison or on probation (Butterfield, 1995). The U.S. prison population has tripled in the last fifteen years (Zachary, 1995).

THE CURRICULUM CHALLENGES AHEAD

By the 21st century, it will be clear that a strategy of decentralization in the area of curriculum will not be working as a tactic to enable the United States to remain competitive in a global economy. Leaving curriculum decisions to 15,000 school boards produces a polyglot curriculum geared to the lowest common denominators of pupil achievement. Exacting standards will not be set locally. If they are, they will most likely be mediocre.

Without national standards of learning, U.S. students will still be the victims of grade inflation as parents continue to put pressure on schools to give high grades for college entrance purposes (see McAdams, 1993) instead of insisting on rigorous achievement as the index to high grades. A decentralized approach to decision making can work only insofar as there is a consensus about national curriculum academic indices. Coming into the 21st century, there is not only no consensus about such benchmarks, there is political antipathy toward defining national standards at all.

Educationally, the 21st century will begin badly for America. The disparity between achievement by socioeconomic levels will widen, funding will not be substantially improved, and the curriculum will be in a state of disarray. Despite decades of experimentation and tinkering, inner-city schools will still be floundering, and test scores will still show pupil learning at the bottom of international comparisons. Perhaps in the early years of the 21st century, it can be clearly shown that promises of reform and change have not been kept. Limited privatization schemes will be shown to have little effect in producing any large-scale change or improvement. We believe that things will have to get much worse before they get better in order to impact a critical mass of people to rethink the existing set of sacred cows that stand as barriers to substantive curriculum improvement. Here are the trends we predict most likely to occur in the early decades of the 21st century:

1. There will be a reemergence of the federal presence in education. Education will be declared in a state of national emergency. New powers will be voted by Congress to set national standards, develop national tests, and improve public education in the inner cities and rural areas of America. The reemerging emphasis on centralized initiatives will be buttressed by Supreme Court rulings that call on the reduction of state-by-state inequities in funding education. New international agencies will be created to provide an accurate data base to the legislative authorities of the nations of the world. Some of these will be supported by global corporations that depend on an educated, multi-national work force.
2. The power of local school boards will be reduced. Already on the wane, the power of local boards of education to engage in curricular change will be eliminated or maintained at the symbolic level only. Sweeping curricular changes will be federally created, with the states ensuring a more equitable tax base for funding schools. Funding will be monitored nationally as guidelines will be set for acceptable deviations in expenditures.
3. National standards for teacher preparation and licensure will become a reality. Teachers will be prepared by universities, but licensed nationally. There will be one common re-

tirement fund, which connects the respective states and enables teachers in short supply to move easily to areas where they are needed. Teachers will be thoroughly prepared in curriculum national standards and be considered the "watchdogs" of curricular focus and connectivity.

4. Corporate involvement in education will expand and corporations will run whole school systems. Entire corporations will operate their own school systems, employing hundreds of teachers, pre-K–16, particularly in areas where good public schools are not available. Such centers will become the place where innovative practices first appear and can be studied appropriately. Corporate centers will become like lab schools, demonstrating new ideas in education and speeding up schools' capacities to undergo and sustain change.

5. Curriculum will become more globally centered and less locally bound. Quaint customs for curriculum, such as learning a respective city's history or a state's history as mandated curriculum, will lose ground to international concerns about the environment, international cooperation, and the need for a common world perspective on world problems without sacrificing cultural diversity and replacing it with cultural hegemony.

6. Pupil assessment will be broadened to include the ability to think critically, solve problems, and question values. Despite protestation by conservative groups, pupil assessment will be dramatically expanded to include the capability to think, deal with ambiguous situations and values, and engage in problem solving. Not every culture of the world embraces teaching its young to think critically, but an interdependent, scientific world will place a premium upon these types of capabilities and outcomes. International assessment will reflect these trends.

CLOSING THOUGHT

The challenge of reconceptualizing what American public school education should be lies ahead of us. Tinkering with the present model is insufficient for developing a learning place where all children learn at high levels. The difficulty of the task is compounded by the persistence of local autonomy, union contracts, the age of the current teaching cadre, and students' lack of high expectations for learning. Nevertheless, setting internationally benchmarked content and performance assessment standards and providing students with a learning environment that enables them to achieve these standards will play a key role in the transformation. As standards are better articulated and supports for high-quality products and performances are put in place, both teachers and students raise their sights (Darling-Hammond, Ancess, and Falk, 1995, p. 260). Perhaps by the year 2025, we will have created such a learning environment.

REFERENCES

Biers, D. (1995, October 24). Asian countries aim to boost research. *Wall Street Journal,* A19.

Bradsher, K. (1995, October 27). Widest gap in incomes? Research points to the U.S. *New York Times,* C2.

Butterfield, F. (1995, October 5). More blacks in their 20's have trouble with the law. *New York Times,* A8.

Darling-Hammond, Linda, Ancess, Jacqueline, and Falk, Beverly. (1995). *Authentic assessment in action.* New York: Teachers College Press.

Kahn, J. (1995, November 13). China swiftly becomes an exporting colossus, straining western ties. *Wall Street Journal,* A1.

McAdams, R. (1993). *Lessons from abroad.* Lancaster, PA: Technomic Publishing Co., Inc.

The family. (1995, September 9). *The Economist.* 25–29.

The world's 100 largest banks (1995, October 2). *The Wall Street Journal,* R33.

Wessel, D. (1995, September 18). World bank ranking of nations' wealth puts Australia on top, the U.S. 12th. *The Wall Street Journal,* R33.

Zachary, G. P. (1995, September 29). Economists say prison boom will take toll. *The Wall Street Journal.* B1.

INDEX

A Nation at Risk, 125
Ability grouping, 27
Academic instructional time, 44
Accountability, 10, 68
Accuracy standards, 120
Achievement, 8, 70–71
ADLER, M., 17
Alignment, 1, 11, 29, 52
American Federation of Teachers, 50
ANCESS, J., 64, 84–85, 92, 95, 141
APPLE, M., 15, 27–28, 124
Apparatus, 23
Applied learning, 62
ARNOVE, R., 31
ARONOWITZ, S., 3
Articulation, 2
Asian Century, 137
Assessment practices, 67
Auditing, 5
Authentic assessment, 72–74
AYER, R., 23

Backloading, practice of, 3
BARTON, P., 67
Behaviorism, 15
Bell-shaped curve, 11
BERNSTEIN, B., 32
BIERS, D., 137
BLANK, R., 132
BOWLES, S., 6
BRADSHER, K., 6
BRANDON, D., 6

Bronx New School, 92
BUTTERFIELD, F., 138

CALLAHAN, R., 124
CAREY, R., 108
CETRON, M., 131
China, 137
CHOMSKY, N., 16
CHUBB, J., 124
Classroom life, 5
COLEMAN, J., 6
COLEY, R., 67
Collaboration, 86
Communitive skills, 60
Competitiveness, 31
Connectivity, 2
Constructivism, 16
Context analysis, 103
Continuous improvement, 128
Coordination, 2
Corporate culture, 124
Correspondence theory, 6
Cultural capital, 6
Cultural code, 32
Cultural literacy, 18
Cultural wars, 24
Curriculum content, 27
Curriculum core, 16
Curriculum delivery, 7
Curriculum design, 7
Curriculum development, 2
Curriculum format, 27

143

Curriculum guides, 28
Curriculum hole, 13
Curriculum management, 4
Curriculum mapping, 12–13
Curriculum metaphors, 14–15
Curriculum planning, 24
Curriculum standards, 37
CURRY, B., 45, 64

DARLING-HAMMOND, L., 64, 84–85, 92, 95, 141
DARDER, A., 22
DELANDSHERE, G., 67
De-skilling, of teachers, 28
Distance learning, 60
DLUGOSH, L., 126

Education Summit, 41
Educational Testing Service, 84
EISNER, E., 16
Emancipatory practices, 22
ENGLISH, F., 3–6, 11–12, 18, 21–22, 134
Evaluation, 30
Evaluation designs, 105
Exam content, 51
Expressive objective, 16

Fair assessment, 102
Feasibility standards, 98
Final vocabularies, 25–26
FINE, M., 124
FINK, A., 97, 117
FINN, C., 124
FOUCAULT, M., 22
France, 76
Frontloading, practice of, 3
FUENTES, B., 32
FULK, B., 84–85, 92–95, 141

GALBRAITH, J., 132
GANDAL, M., 35–39
GANDHI, M., 23
GARCIA, R., 32
Germany, 49

GIBBONEY, R., 127
GINTIS, H., 6
GIROUX, H., 3, 21–24, 125
GLATTHORN, A., 19
Global economy, 36
Goals 2000, 38–40
GONZALES, R., 32
GORMAN, J., 35–36
GREENE, S., 133
GUSKEY, T., 126
GUTIERREZ, K., 24–27

HAERTEL, G., 68
Hegemonic curriculum, 21
HERMAN, J., 97, 105, 114
HERRNSTEIN, R., 125
Hidden curriculum, 3, 21
HILL, J., 3, 22
HIRSCH, E., 18
Hispanic population, 43
HOUSE, E., 130
HOWE, H., 123
HUNTER, M., 126

Ideal vision, 26
Instruction, 1
International high school, 85
International standards, 129

J curve, 11

KAESTLE, C., 126
KAUFMAN, R., 26
Kentucky Reform, 42–43
KINCHELOE, J., 24
Knowledge, production of, 22, 67
KOBUS, M., 60

LABELLE, T., 29
LARSON, J., 24–27
Learning, 8
LETENDRE, G., 32
LEWIS, A., 37–64
Lived curriculum, 3
Loose coupling, 4

MAGNER, D., 123
Management by objectives, 132
MANATT, R., 23–25, 126
MCADAMS, R., 138
MCLAREN, P., 22–24, 27
MEIER, D., 75
Melting pot, 32
Meta evaluation, 107
MOE, T., 124
Monocultural curriculum cores, 32
Morals, 23
MULLIGAN, C., 31

NADLER, D., 132
NAISBETT, J., 131
Nation building, 31
National Assessment of
    Educational Progress, 5
National Defense Education Act,
    123–124
National Education Goals Panel, 38
National standards, 36, 62
New Standards Project, 63
NOWAKOWSKI, J., 4

Operational indicators, 26
Open-ended response items, 76–79
Opportunity to learn (OTL), 51
Outcomes based education, 45

PAGE, J., 34
PAGE, R., 21–27
Paradigm shift, 59
Performance event, 80
PETROSKY, A., 67
PIAGET, J., 16
PITMAN, M., 18
Poor schools, 10
Portfolio assessment, 71–72
POSAVAC, E., 108
Poverty, 124–125
Privatization, 130
Problem solving skills, 60
Program evaluation, 97

QUINN, D., 4

Randomness, 11
RATTNER, S., 6
Research skills, 60
Reliability, 68
Resistance, to evaluation, 108
ROBERTS, R., 125
ROBINSON, G., 6
ROGERS, C., 17
RORTY, R., 26

SCANS Report, 63
Scheduling practices, 27
School, 3
School boards, 126
School reform, 84
Scientific literacy, 129
Seamless web, 8–9
Segregation, 18
SHORT, I., 24
SINGER, I., 125
SKINNER, B., 15
Slam bang effect, 108
SMITH, M., 45
SPRING, J., 124–127
Stakeholders, 99
Standards, types of, 98
STANLEY, W., 21
State assessment practices, 68–70
State education agencies, 11
State reproduction model, 22
STEFFY, B., 7, 12, 53, 65–67
STEWART, M., 4
Strategic plan, 131
Structure, 3
Structured silences, 3
Student inquiry, 75
Sub-optimization, 134
Summative evaluation, 107
Socio-economic determinism, 5
Socio-economic status, 5

Taught curriculum, 12
Teacher, 61

Teacher/student relationship, 61
Teaching to the test, 9–11
Teamwork, 60
Technobabble, 24
Technology, 62
TEMPLE, T., 45
Test scores, 7
THORNDIKE, E., 12
Transfer theory, 12
TIVNAN, E., 25

Unaligned curriculum, 11
Utility standards, 98

Validity, 68
Vouchers, 17, 130

WAGNER, R., 126
WATSON, C., 132
WEICK, K., 4
WESSEL, D., 138
WESTBURY, I., 130
WHEELER, P., 68
Whole language, 16
Work plan, 1
World class standards, 43

ZACHARY, G., 138